Praise for
This Is Your Mind on Plants

"Delightful . . . [*This Is Your Mind On Plants*] aims to collapse the distinctions between legal and illegal, medical and recreational, exotic and everyday, by appealing to the principle that unites the three: the affinities between plant biochemistry and the human mind."
—*The New York Review of Books*

"[A] thoughtful study . . . As the U.S.'s drug policies become less punitive, [Pollan] argues, we should think more clearly about substances we've come to depend on."
—*The New Yorker*

"[A] wonderful and compelling read that will leave you thinking long after you set it down. . . . Pollan is an astonishingly good writer, at times intimate and vulnerable, at times curious and expository, always compelling and credible. Reading his writing can be kind of like taking a psychedelic—a literary onomatopoeia." —*The Washington Post*

"Pollan is a mindful and enthusiastic psychonaut. He is also a gifted writer who synthesizes unruly social histories and wreathes them around his own drug-taking experiences. And he articulates these experiences with great insight and eloquence." —*The New Republic*

"Expert storytelling . . . [Pollan] masterfully elevates a series of big questions about drugs, plants, and humans that are likely to leave readers thinking in new ways."
—*The New York Times Book Review*

"Fascinating . . . *This Is Your Mind on Plants* has much to offer its readers, whether they are curious about the plant-based adventures of others or the science of substances at work in their own minds. With historical depth, political punch, and narrative exuberance, Pollan's book sounds a call to reimagine society's relationship with psychoactive plants."
—*The Boston Globe*

"Pollan's insatiable appetite to learn every possible morsel about the subject on which he is writing is a gift that has proved itself with bestseller after bestseller. . . . Anchored by a refreshing willingness to expose his own blind spots, [*This Is Your Mind on Plants*] is an engrossing, plant-powered blend of general history, contemporary reporting, and potent self-reflection." —*San Francisco Chronicle*

"The author of *How to Change Your Mind* turns his attention to three consciousness-altering drugs—opium, mescaline, and caffeine (yes, it's a drug)—in this eye-opening exploration." —*People*

"Highly engaging reading . . . Pollan's writing always has a personal aspect to it, but in his latest work he takes an even more central role in the narrative, and his book is the better for it." —*The Daily Beast*

"Pollan weaves together three separately engaging stories in a pleasantly meandering style, deftly using his personal experiences with each compound as a jumping-off point for small forays into anthropology, history, politics, psychology, molecular biology, and neuroscience. Even the most distracted reader will come away with an understanding of the physical effects of the spotlighted substances as well as their cultural significance."

—*Science*

"The omnivorously curious Pollan pivots off his provocative *How to Change Your Mind* with an enthralling odyssey into a trio of mind-altering drugs found in plants: opium, caffeine, and mescaline. In this wide-ranging, deliciously written study, he asks, why does one power us up each morning while the other two are shrouded in taboo? You'll never look at a Starbucks Pike's Peak the same way again." —*Oprah Daily*

"This fascinating insight into our relationship with mind-altering plants weaves personal experimentation with cultural history. . . . Pollan is always an entertaining writer, and a deep thinker with a light touch . . . curious, careful, open-minded."

—*The Guardian* (London)

"Like it or not, we are undergoing a drugs revolution, slowly stripping back the inhibitions of the mid-twentieth century . . . thankfully Pollan is here to guide us through this putative challenge." —*The Sunday Times (London)*

"A brilliantly objective examination of our relationships with opium, caffeine, and mescaline . . . The descriptions of London's coffeehouse culture and Honoré de Balzac's barbarous habit of ingesting dry coffee grounds to fuel all-night scribbling sessions are worth the book's price alone. . . . The book is really about the relation between each plant and the humans who consume it, tackled in a nonjudgmental and objective way that seeks to dispel the ignorance, prejudice, and demonization they attract." —*Financial Times*

"*This Is Your Mind on Plants* is witty, entertaining, and polite, but it is not trivial. Subtly but assuredly, Pollan argues that which plants (and fungi) we are allowed and how depends, consciously or otherwise, on the interests of power."

—*The Times Literary Supplement* (London)

"Fascinating and occasionally terrifying . . . Pollan has long been interested in the things we put in our bodies, and here he studies three drugs derived from plants that alter human consciousness: opium, caffeine, and mescaline. . . . His opium chapter is mesmerizing . . . entertaining." —*Daily Mail (London)*

"*This Is Your Mind on Plants* is an entertaining blend of memoir, history, and social commentary that illustrates Pollan's ability to be both scientific and personal. By relying on contextual history and focusing on three popular, if misunderstood, drugs, Pollan challenges common views on what mind-altering drugs are and what they can accomplish."

—*BookPage* (starred review)

"Pollan is a master of breaking down complex science into an engaging story and challenging long-held societal beliefs. His newest offering, which follows his examination of the science of psychedelics in 2018's *How to Change Your Mind*, aims to unpack our ideas about what constitutes a 'drug' and, fundamentally, why we seek them." —*Time*

"Building on his lysergically drenched book *How to Change Your Mind* (2018), Pollan looks at three plant-based drugs and the mental effects they can produce. . . . A lucid (in the sky with diamonds) look at the hows, whys, and occasional demerits of altering one's mind." —*Kirkus Reviews* (starred review)

PENGUIN BOOKS

THIS IS YOUR MIND ON PLANTS

Michael Pollan is the author of nine books, including *How to Change Your Mind*, *Cooked*, *Food Rules*, *In Defense of Food*, *The Omnivore's Dilemma*, and *The Botany of Desire*, all of which were *New York Times* bestsellers. He is also the author of the audiobook *Caffeine: How Coffee and Tea Created the Modern World*. A longtime contributor to *The New York Times Magazine*, Pollan teaches writing at Harvard University and the University of California, Berkeley. In 2010, *Time* magazine named him one of the one hundred most influential people in the world.

ALSO BY MICHAEL POLLAN

How to Change Your Mind

Cooked

Food Rules

In Defense of Food

The Omnivore's Dilemma

The Botany of Desire

A Place of My Own

Second Nature

THIS IS YOUR MIND on PLANTS

Michael Pollan

PENGUIN BOOKS

PENGUIN BOOKS
An imprint of Penguin Random House LLC
penguinrandomhouse.com

First published in the United States of America by Penguin Press,
an imprint of Penguin Random House LLC, 2021
Published in Penguin Books 2022

"Opium, Made Easy" was originally published in *Harper's Magazine*, April 1, 1997.

Caffeine: How Coffee and Tea Created the Modern World was originally released as an audiobook by Audible Originals, January 30, 2020.

ISBN 9780593296929 (paperback)
ISBN 9780593493519 (international paperback)

THE LIBRARY OF CONGRESS HAS CATALOGED THE HARDCOVER EDITION AS FOLLOWS:
Names: Pollan, Michael, author.
Title: This is your mind on plants / Michael Pollan.
Description: New York: Penguin Press, 2021. |
Includes bibliographical references and index.
Identifiers: LCCN 2021003519 (print) | LCCN 2021003520 (ebook) |
ISBN 9780593296905 (hardcover) | ISBN 9780593296912 (ebook)
Subjects: LCSH: Psychotropic plants. | Opium. | Mescaline. | Caffeine.
Classification: LCC RS164 .P764 2021 (print) |
LCC RS164 (ebook) | DDC 581.6—dc23
LC record available at https://lccn.loc.gov/2021003519
LC ebook record available at https://lccn.loc.gov/2021003520

Printed in the United States of America
3 5 7 9 10 8 6 4

Designed by Amanda Dewey

For Judith,

for sharing the journey

CONTENTS

THIS IS YOUR MIND ON PLANTS

INTRODUCTION

Of all the many things humans rely on plants for—sustenance, beauty, medicine, fragrance, flavor, fiber—surely the most curious is our use of them to change consciousness: to stimulate or calm, to fiddle with or completely alter, the qualities of our mental experience. Like most people, I use a couple of plants this way on a daily basis. Every morning without fail I begin my day by preparing a hot-water infusion of one of two plants that I depend on (and dependent I am) to clear the mental fog, sharpen my focus, and prepare myself for the day ahead. We don't usually think of caffeine as a drug, or our daily use of it as an addiction, but that is only because coffee and tea are legal and our dependence on them is socially acceptable. So, then, what exactly is a drug? And why is making tea from the leaves of *Camellia sinensis* uncontroversial, while doing the same thing with the seed heads of *Papaver somniferum* is, as I discovered to my peril, a federal crime?

All who try to construct a sturdy definition of drugs eventually run aground. Is chicken soup a drug? What about sugar? Artificial sweeteners? Chamomile tea? How about a placebo? If we define a drug simply as a substance we ingest that changes us in some way, whether in body or in mind (or both), then all those substances surely qualify. But shouldn't we be able to distinguish foods from drugs? Faced with that very dilemma, the Food and Drug Administration punted, offering a circular definition of drugs as "articles other than food" that are recognized in the pharmacopoeia—that is, as drugs by the FDA. Not much help there.

Things become only slightly clearer when the modifier "illicit" is added: an illicit drug is whatever a government decides it is. It can be no accident that these are almost exclusively the ones with the power to change consciousness. Or, perhaps I should say, with the power to change consciousness in ways that run counter to the smooth operations of society and the interests of the powers that be. As an example, coffee and tea, which have amply demonstrated their value to capitalism in many ways, not least by making us more efficient workers, are in no danger of prohibition, while psychedelics—which are no more toxic than caffeine and considerably less addictive—have been regarded, at least in the West since the mid-1960s, as a threat to social norms and institutions.

But even these classifications are not as fixed or as sturdy as you may think. At various times both in the Arab world and in Europe, authorities have outlawed coffee, because they regarded the people who gathered to drink it as politically threatening. As I write, psychedelics seem to be undergoing a change of identity. Since researchers have demonstrated that psilocybin can be useful in treating mental health, some psychedelics will probably soon become FDA-approved medicines: that is, recognized as more helpful than threatening to the functioning of society.

This happens to be precisely how Indigenous peoples have always regarded these substances. In many Indigenous communities, the ceremonial use of peyote, a psychedelic, *reinforces* social norms by bringing people together to help heal the traumas of colonialism and dispossession. The government recognizes the First Amendment right of Native Americans to ingest peyote as part of the free exercise of their religion, but under no circumstances do the rest of us enjoy that right, even if we use peyote in a similar way. So here is a case where it is the identity of the user rather than the drug that changes its legal status.

Nothing about drugs is straightforward. But it's not quite true that our plant taboos are entirely arbitrary. As these examples suggest, societies condone the mind-changing drugs that help uphold society's rule and ban the ones that are seen to undermine it. That's why in a society's choice of psychoactive substances we can read a great deal about both its fears and its desires.

Ever since I took up gardening as a teenager and attempted to grow cannabis, I have been fascinated by our attraction to these powerful plants as well as by the equally powerful taboos and fraught feelings with which we surround them. I've come to appreciate that when we take these plants into our bodies and let them change our minds, we are engaging with nature in one of the most profound ways possible.

There is scarcely a culture on earth that hasn't discovered in its environment at least one such plant or fungus, and in most cases a whole suite of them, that alters consciousness in one of a variety of ways. Through what was surely a long and perilous trial and error, humans have identified plants that lift the burden of physical pain;

render us more alert or capable of uncommon feats; make us more sociable; elicit feelings of awe or ecstasy; nourish our imagination; transcend space and time; occasion dreams and visions and mystical experiences; and bring us into the presence of our ancestors or gods. Evidently, normal everyday consciousness is not enough for us humans; we seek to vary, intensify, and sometimes transcend it, and we have identified a whole collection of molecules in nature that allow us to do that.

This Is Your Mind on Plants is a personal inquiry into three of those molecules and the remarkable plants that produce them: the morphine in the opium poppy; the caffeine in coffee and tea; and the mescaline produced by the peyote and San Pedro cacti. The second of these molecules is legal everywhere today; the first is illegal in most places (unless it has been refined by a pharmaceutical company and prescribed by a physician); and the third is illegal in the United States unless you are a member of a Native American tribe. Each represents one of the three broad categories of psychoactive compounds: the downer (opium); the upper (caffeine); and what I think of as the outer (mescaline). Or, to put it a bit more scientifically, I profile here a sedative, a stimulant, and a hallucinogen.

Taken together, these three plant drugs cover much of the spectrum of the human experience of psychoactive substances, from the everyday use of caffeine, the most popular psychoactive drug on the planet; to the ceremonial use of mescaline by Indigenous peoples; to the age-old use of opiates to relieve pain. That particular chapter is set during the drug war, at a topsy-turvy moment when the government was paying more attention to a bunch of gardeners growing poppies in order to brew a mild narcotic tea than it was to a pharmaceutical company that was knowingly addicting millions of

Americans to its FDA-approved opiate, OxyContin. I was one of those gardeners.

I tell each of these stories from multiple perspectives and through a variety of lenses: historical, anthropological, biochemical, botanical, and personal. In each case, I have some skin in the game—or perhaps I should say brain cells, since I don't know how to write about how it feels, and what it means, to change consciousness without conducting some self-experimentation. Though in the case of caffeine, self-experimentation meant abstaining from it rather than partaking, which proved much harder to do.

One of these chapters consists of an essay I wrote twenty-five years ago, when the drug war was raging, and it bears the scars of that period of fear and paranoia. But the other stories have been inflected by the fading of that war, the end of which now appears in sight. In the 2020 election, Oregonians voted to decriminalize the possession of *all* drugs and specifically to legalize therapy using psilocybin. A ballot measure passed in Washington, D.C., calls for the decriminalization* of "entheogenic plants and fungi." ("Entheogen," from the Greek for "manifesting the god [divine] within," is an alternative term for psychedelics, coined in 1979 by a group of religious scholars hoping to remove the counterculture taint from this class of drugs and underscore the spiritual use to which they have been put for thousands of years.) In the same election, New Jersey, along with four traditionally red states—Arizona, Mississippi, Montana, South Dakota—voted to liberalize marijuana laws, bringing

*"Decriminalize" is a bit of a misnomer; the ballot measure instructs law enforcement and prosecutors to make the prosecution of crimes involving the growing, possession, or use—but not the sale—of plant medicines their lowest priority. The campaign was organized by a new drug-reform movement called Decriminalize Nature, which I discuss in the chapter on mescaline.

the number of states that have legalized some form of marijuana use to thirty-six.

My wager in writing *This Is Your Mind on Plants* is that the decline of the drug war, with its brutally simplistic narratives about "your brain on drugs," has opened a space in which we can tell some other, much more interesting stories about our ancient relationship with the mind-altering plants and fungi with which nature has blessed us.

I use the word "blessed" in full awareness of the human tragedies that can accompany the use of drugs. Much better than we do, the Greeks understood the two-faced nature of drugs, an understanding reflected in the ambiguity of their term for them: *pharmakon*. A pharmakon can be either a medicine or a poison; it all depends—on use, dose, intention, and set and setting.* (The word has a third meaning as well, one often relied on during the drug war: a pharmakon is also a scapegoat, something for a group to blame its problems on.) Drug abuse is certainly real, but it is less a matter of breaking the law than of falling into an unhealthy relationship with a substance, whether licit or illicit, one in which the ally, or medicine, has become an enemy. The same opiates that killed some fifty thousand Americans by overdose in 2019 also make surgery endurable and ease the passage out of this life. Surely that qualifies as a blessing.

The stories I tell here put this trio of psychoactive plant chemicals into the context of our larger relationship to nature. One of the innumerable threads connecting us to the natural world is the one

*"Set and setting" is the term Timothy Leary introduced to underscore the powerful influence of one's mind-set and physical setting in shaping a psychedelic experience.

that links plant chemistries to human consciousness. And since this *is* a relationship, we need to account for the plants' points of view as well as our own. How amazing is it that so many kinds of plants have hit upon the precise recipes for molecules that fit snugly into receptors in human brains? And that by doing so these molecules can short-circuit our experience of pain, or rouse us, or obliterate the sense of being a separate self? You have to wonder: what's in it for the plants to devise and manufacture molecules that can pass for human neurotransmitters and affect us in such profound ways?

Most of the molecules that plants produce that change animal minds start out as tools for defense: alkaloids like morphine, caffeine, and mescaline are bitter-tasting toxins meant to discourage animals from eating the plants that make them and, should the animals persist, to poison them. But plants are clever, and over the course of evolution they've learned that simply killing a pest outright is not necessarily the smartest strategy. Since a lethal pesticide would quickly select for resistant members of the pest population, rendering it ineffective, plants have evolved subtler and more devious strategies: chemicals that instead mess with the minds of animals, confusing or disorienting them or ruining their appetite—something that caffeine, mescaline, and morphine all reliably do.

But while most of the psychoactive molecules plants have developed started out as poisons, they sometimes evolved into the opposite: attractants. Scientists recently discovered a handful of species that produce caffeine in their nectar, which is the last place you would expect a plant to serve up a poisonous beverage. These plants have discovered that they can attract pollinators by offering them a small shot of caffeine; even better, that caffeine has been shown to sharpen the memories of bees, making them more faithful, efficient, and hardworking pollinators. Pretty much what caffeine does for us.

Once humans discovered what caffeine and morphine and mescaline could do for them, the plants that produce the greatest amounts of these chemicals were the ones that prospered in the sunshine of our attention; we disseminated their genes around the world, vastly expanding their habitat and providing for their every need. By now our fates and the fates of these plants are complexly intertwined. What began as war has evolved into marriage.

Why do we humans go to such lengths to change our minds, and then why do we fence that universal desire with laws and customs, taboos and anxieties? These questions have occupied me since I began writing about our engagement with the natural world more than thirty years ago. When you compare this desire to the other needs we turn to nature to gratify—for food, clothing, shelter, beauty, and so on—the drive to alter consciousness wouldn't seem to contribute nearly as much, if anything, to our success or survival. In fact, the desire to change consciousness may be seen as maladaptive, since altered states can put us at risk for accidents or make us more vulnerable to attack. Also, many of these plant chemicals are toxic; others, like morphine, are highly addictive.

But if our species' desire to change consciousness is universal, a human given, then doing so should offer benefits to make up for the risks, or natural selection would long ago have weeded out the drug takers. Take, for example, morphine's value as a painkiller, which has made it one of the most important drugs in the pharmacopoeia going back thousands of years.

Plants that change consciousness answer to other human needs as well. We shouldn't underestimate the value, to people trapped in monotonous lives, of a substance that can relieve boredom and entertain by sponsoring novel sensations and thoughts in the mind. Some drugs can expand the contours of a world constrained by circumstance, as I discovered during the pandemic. Drugs that enhance sociability not only gratify us but presumably result in more offspring. Stimulants like caffeine improve concentration, making us better able to learn and work, and to think in rational, linear ways. Human consciousness is always at risk of getting stuck, sending the mind around and around in loops of rumination; mushroom chemicals like psilocybin can nudge us out of those grooves, loosening stuck brains and making possible fresh patterns of thought.

Psychedelic drugs can also benefit us—and occasionally our culture—by stimulating the imagination and nourishing creativity in the individuals who take them. This is not to suggest that all the ideas that occur to the altered mind are any good; most of them aren't. But every now and then a tripping brain will hit upon a novel idea, a solution to a problem, or a new way of looking at things that will benefit the group and, possibly, change the course of history. The case can be made that the introduction of caffeine to Europe in the seventeenth century fostered a new, more rational (and sober) way of thinking that helped give rise to the age of reason and the Enlightenment.

It's useful to think of these psychoactive molecules as mutagens, but mutagens operating in the realm of human culture rather than in biology. In the same way that exposure to a disruptive force like radiation can mutate genes, introducing variation and throwing off new traits that every so often prove adaptive for the species, psychoactive drugs, operating on the minds of individuals, occasionally contribute useful new memes to the evolution of culture—conceptual breakthroughs, fresh metaphors, novel theories. Not always, not even often, but every now and then, the encounter of a mind and a plant molecule changes things. If the human imagination has a natural history, as it must, can there be any doubt that plant chemistries have helped to inform it?

Psychedelic compounds can promote experiences of awe and mystical connection that nurture the spiritual impulse of human beings—indeed, that might have given rise to it in the first place, according to some religious scholars.* The notion of a beyond, of a hidden dimension of reality, or of an afterlife—these, too, may be memes introduced to human culture by visions that psychoactive molecules inspired in human minds. Drugs are not the only way to occasion the sort of mystical experience at the core of many religious traditions—meditation, fasting, and solitude can achieve similar results—but they are a proven tool for making it happen. The spiritual or ceremonial use of plant drugs can also help knit people together, fostering a stronger sense of social connection accompanied by a diminished sense of self. We have only just begun to understand how the human involvement with psychoactive plants has shaped our history.

---

*The idea that psychedelics have played a foundational role in religion has been floating around the fringes of religious studies since at least the 1970s, when R. Gordon Wasson (the man who rediscovered psilocybin) collaborated with Albert Hofmann (the inventor of lysergic acid diethylamide, or LSD) and a young classicist named Carl A. P. Ruck to write *The Road to Eleusis: Unveiling the Secret of the Mysteries* (New York: Harcourt Brace Jovanovich, 1978; reprint, Berkeley: North Atlantic Books, 2008). See also John M. Allegro, *The Sacred Mushroom and the Cross* (London: Hodder and Stoughton; New York: Doubleday, 1970). An excellent recent exploration of the role of psychedelics in early religion is found in Brian C. Muraresku's *The Immortality Key: The Secret History of the Religion with No Name* (New York: St. Martin's Press, 2020).

It probably shouldn't surprise us that plants of such power and possibility are surrounded by equally powerful emotions, laws, rituals, and taboos. These reflect the understanding that changing minds can be disruptive to both individuals and societies, and that when such powerful tools are placed in the hands of fallible human beings, things can go very wrong. We have much to learn from traditional Indigenous cultures that have made long use of psychedelics like mescaline or ayahuasca: as a rule, the substances are never used casually, but always with intention, surrounded by ritual and under the watchful eye of experienced elders. These people recognize that these plants can unleash Dionysian energies that can get out of control if not managed with care.

But the blunt instrument of a drug war has kept us from reckoning with these ambiguities and the important questions about our nature that they raise. The drug war's simplistic account of what drugs do and are, as well as its insistence on lumping them all together under a single meaningless rubric, has for too long prevented us from thinking clearly about the meaning and potential of these very different substances. The legal status of this or that molecule is one of the least interesting things about it. Much like a food, a psychoactive drug is not a thing—without a human brain, it is inert—so much as it is a relationship; it takes both a molecule and a mind to make anything happen. The premise of this book is that these three relationships hold up mirrors to our deepest human needs and aspirations, the operations of our minds, and our entanglement with the natural world.

OPIUM

Prologue

The narrative that follows this prologue is something of a period piece, a dispatch from the war on drugs near its peak, circa 1996–97, that itself became a minor casualty of that war. The piece originally appeared in the April 1997 issue of *Harper's Magazine*, but not in its entirety. After consulting with several lawyers, I concluded there were four or five crucial pages of the narrative that I couldn't publish without risking arrest as well as the forfeiture of our house and garden—the wrecking of our life, basically. Twenty-four years later, those pages—which had gone missing after I hid them away—have been restored and appear here in print for the first time.

The story began as something of a lark and ended in anxiety, paranoia, and self-censorship. At the time, my wife and I and our four-year-old son were living in rural Connecticut, and I was writing personal essays about the goings-on in my garden. As a gardener, I'd become fascinated by the symbiotic relationship our species has struck up with certain plants, using them to gratify our desires for everything from nourishment to beauty to a change of consciousness. Early in 1996, my editor at *Harper's Magazine*, Paul Tough, sent me an underground-press book called *Opium for the Masses* that had crossed his desk, suggesting there might be a column in it for me. I was immediately intrigued by the idea that I could grow opium and produce this most ancient of psychoactive drugs in my garden from easily obtainable seeds. I decided to give it a try, just to see what would happen. What happened turned out to be a living nightmare, as I found myself ensnared in a quiet but determined federal campaign to stamp out knowledge of an easy-to-produce homegrown narcotic before it became a fad.

Read today, in what we can hope are the waning days of the drug war, the piece feels overwrought in places, but it's important to understand the context in which it was written. Under President Clinton, the government was prosecuting the drug war with a vehemence never before seen in America. The year I planted my poppies, more than a million Americans were arrested for drug crimes. The penalties for many of those crimes had become draconian under Clinton's 1994 crime bill, which introduced new "three-strikes" sentencing provisions and led to mandatory minimum sentences for many nonviolent drug offenses. By the mid-1990s, a series of Supreme Court decisions in drug cases had handed the government a raft of new powers that have significantly eroded our civil liberties. The government also won new powers to confiscate property—houses, cars, land—involved in drug crimes, even when no individual has been convicted, or even charged.

Were these erosions of our liberties a casualty of the drug war or its objective? It's a fair question. President Clinton didn't start the drug war—that distinction belongs to Richard Nixon, who we now know viewed drug enforcement not as a matter of public health or safety but as a political tool to wield against his enemies. In an April 2016 article in *Harper's Magazine*, "Legalize It All," Dan Baum recounted an interview that he conducted with John Ehrlichman in 1994—two years before my misadventures in the garden. Ehrlichman, you will recall, was President Nixon's domestic policy adviser; he served time in federal prison for his role in Watergate. Baum came to talk to Ehrlichman about the drug war, of which he was a key architect.

"You want to know what this was really all about?" Ehrlichman began, startling the journalist with both his candor and his cynicism. Ehrlichman explained that the Nixon White House "had two enemies: the antiwar left and black people. . . . We knew we couldn't make it illegal to be either against the war or black, but by getting the public to associate the hippies with marijuana and blacks with heroin, and then criminalizing both heavily, we could disrupt those communities. We could arrest their leaders, raid their homes, break up their meetings, and vilify them night after night on the evening news. Did we know we were lying about the drugs? Of course we did."*

Although neither victory nor defeat was ever declared in the war on drugs, you seldom hear the phrase on the lips of government officials and politicians anymore. I suspect there are two reasons for their silence: As a matter of politics, the government has less need of draconian drug laws since declaring a new "war" in 2001. The war on terror has taken over from the war on drugs as a justification for expanding government power and curbing civil liberties. And as a matter of public health, it has become obvious to anyone paying attention that, after a half century of waging war on drugs, it is the drugs

*The quote has been disputed by some of Ehrlichman's colleagues in the administration; Baum died in 2020, so I was not able to ask him for documentation or an explanation of why he waited more than a decade to publish it.

that are winning. Criminalizing drugs has done little to discourage their use or to lower rates of addiction and death from overdose. The drug war's principal legacy has been to fill our prisons with hundreds of thousands of nonviolent criminals—a great many more of them Black people than hippies. This, then, is the first historical context in which my account of growing opium in 1996 should be read, as a window on a dark and fearful time in America, when you didn't have to leave your garden to become a criminal and put yourself in serious legal jeopardy. But there is another historical context in which the piece can be read, and this one nobody was aware of at the time.

The words "opium" and "opiate" carry a very different set of connotations today than they did when I planted my poppies in 1996. Now the words conjure a national public health catastrophe, but in 1996 there was no "opioid crisis" in America. What there was were maybe half a million heroin addicts, and about forty-seven hundred deaths from drug overdoses each year. At the time, these tragedies were often cited to justify the war on drugs, but in a country of 270 million this hardly qualified as a public health crisis. (Which is the reason cannabis had to be added to the war's list of targets.) Today, by comparison, deaths from overdose of opiates, both licit and illicit, approach fifty thousand a year, and an estimated 2 million Americans are addicted to opiates of one kind or another. (Another 10 million abuse opiates, according to the Substance Abuse and Mental Health Services Administration.) After the coronavirus, the opiate epidemic represents the biggest threat to public health since the AIDS/HIV epidemic.

The chief culprit in the opiate epidemic is not a virus, however, or even the illicit drug economy; it's a corporation. What I didn't know when I was conducting my illegal experiments with opium is that, at the very same historical moment, the pharmaceutical industry was planting the first seeds of the opioid crisis. The same summer that the Drug Enforcement Agency (DEA) was quietly cracking down on gardeners, seed merchants, writers, and other small-timers messing around with opium poppies, a little-known pharmaceutical company called Purdue Pharma—headquartered in Stamford, Connecticut, sixty miles down Route 7 from my garden—had begun marketing a new, slow-release opiate called OxyContin.

Launched in 1996, Purdue's aggressive marketing campaign for OxyContin convinced doctors that the company's new formulation was safer and less addictive than other opiates. The company assured the medical community that pain was being undertreated, and that the new opiate could benefit not just cancer and surgery patients but people suffering from arthritis, back pain, and workplace injuries. The campaign produced an explosion in prescriptions for OxyContin that would earn the company's owners, the Sackler family,* more than $35 billion, while leading to more than 230,000 deaths by overdose. But that figure grossly understates the number of casualties from OxyContin: thousands of people who became addicted to legal painkillers eventually turned to the underground when they could no longer obtain or afford prescription opiates; four out of five new heroin users used prescription painkillers first.

At the same time a war against illicit drugs was raging, ostensibly to stamp out a real but fairly modest public health problem, a legal, FDA-approved opiate was being pushed on people, creating

*The Sacklers joined a tradition of illustrious American families whose fortunes flowed from the sale of opium and its derivatives, including John Jacob Astor and the Cabots, Perkinses, and Cushings of Boston, all much better known for their philanthropy and patronage.

what became a genuine public health crisis. Read in this light, the drug war machinations looming over my garden and story seem almost comic, in a Keystone Kops sort of way. *They went thataway.*

Humans have been cultivating opium poppies for more than five thousand years, as one of the most important medicines in the pharmacopoeia. For most of that time we have recognized the two-faced nature of the flower and the powerful molecules it gives us: that it is at once a blessing—to those in pain or on the verge of death—and a grave peril to any who would abuse it. To both the Greeks and Romans, the poppy flower symbolized both the sweetness of sleep and the prospect of death. We're evidently not as good as they were at holding two contradictory ideas in our heads, for today who has a good word to say about opiates or opium? "Blessing" no longer comes to mind, except perhaps at the deathbed. But what is true of the opium poppy is true for all the medicines that plants have given us: they are both allies and poisons at once, which means it's up to us to devise a healthy relationship with them.

As for the poppy flower itself, it may soon disappear from our age-old relationship to the opiates, as much stronger and cheaper synthetic versions of the flower's alkaloids come to dominate both the legal and illicit markets for painkillers. Something will be lost when that happens. One of the wagers of my experiment in the garden is that there might be some value in getting to know the opium poppy in all its aspects and power, before its role in our lives, once so important, is downgraded to ornament.

Opium, Made Easy

Last season was a strange one in my garden, notable not only for the unseasonably cool and wet weather—the talk of gardeners all over New England—but also for its climate of paranoia. One flower was the cause: a tall, breathtaking poppy, with silky scarlet petals and a black heart, the growing of which, I discovered rather too late, is a felony under state and federal law. Actually, it's not quite as simple as that. My poppies were, or became, felonious; another gardener's might or might not be. The legality of growing opium poppies (whose seeds are sold under many names, including the breadseed poppy, *Papaver paeoniflorum*, and, most significantly, *Papaver somniferum*) is a tangled issue, turning on questions of nomenclature and epistemology that it took me the better part of the summer to sort out. But before I try to explain, let me offer a friendly warning to any gardeners who might wish to continue growing this spectacular annual: the less you know about it, the better off you are, in legal if not horticultural terms. Because whether or not the opium poppies in your garden are illicit depends not on what you do, or even intend to do, with them but very simply on what you *know* about them. Hence my warning: if you have any desire to grow opium poppies, you would be wise to stop reading right now.

As for me, I'm afraid that, at least in the eyes of the law, I'm already lost, having now tasted of the forbidden fruit of poppy knowledge. Indeed, the more I learned about poppies, the guiltier my poppies became—and the more fearful grew my days and to some extent also my nights. Until the day last fall, that is, when I finally pulled out my poppies' withered stalks and, with a tremendous feeling of relief, threw them on the compost, thereby (I hope) rejoining the ranks of gardeners who don't worry about visits from the police.

It started out if not quite innocently, then legally enough. Or at least that's what I thought back in February, when I added a couple of poppy varieties (*P. somniferum* as well as *P. paeoniflorum* and *P. rhoeas*) to my annual order of flowers and vegetables from the seed catalogs. But the state of popular (and even expert) knowledge about poppies is confused, to say the least; mis- and even disinformation is rife. I'd read in *Martha Stewart Living* that "contrary to general belief, there is no federal law against growing *P. somniferum*." Before planting, I consulted my *Taylor's Guide to Annuals*, a generally reliable reference that did allude to the fact that "the juice of the unripe pod yields opium, the production of which is illegal in the United States." But *Taylor's* said nothing worrisome about the plants themselves. I figured that if the seeds could be sold legally (and I found *somniferum* on offer in a half-dozen well-known catalogs, though it was not always sold under that name), how could the obvious next step—i.e., planting the seeds according to the directions on the packet—possibly be a federal offense? Were this the case, you would think there'd at least be a disclaimer in the catalogs.

So it seemed to me that I could remain safely on the sunny side of the law just as long as I didn't attempt to extract any opium from my poppies. Yet I have to confess that this was a temptation I grappled with all last summer. You see, I'd become curious as to whether it was in fact possible, as I'd recently read, for a gardener of average skills to obtain a narcotic from a plant grown in this country from legally available seeds. To another gardener this will not seem odd, for we gardeners are like that: eager to try the improbable, to see if we can't successfully grow an artichoke in Zone 5 or make echinacea tea from the roots of our purple coneflowers. Deep down I suspect that many gardeners regard themselves as minor-league alchemists, transforming the dross of compost (and water and sunlight) into substances of rare value and beauty and power. Also, one of the greatest satisfactions of gardening is the independence it can confer—from the greengrocer, the florist, the pharmacist, and, for some, the drug dealer. One does not have to go all the way "back to the land" to experience the satisfaction of providing for yourself off the grid of the national economy. So, yes, I was curious to know if I could make opium at home, especially if I could do so without making a single illicit purchase. It seemed to me that this would indeed represent a particularly impressive sort of alchemy.

I wasn't at all sure, however, whether I was prepared to go quite that far. I mean, *opium!* I'm not eighteen anymore, or in any position to undertake such a serious risk. I am in fact forty-two, a family man (as they say) and homeowner whose drug-taking days are behind him.* Not that they aren't sometimes fondly recalled, the prevailing cant about drug abuse notwithstanding. But now I have a kid and a mortgage and a Keogh. There is simply no place in my grown-up, middle-class lifestyle for an arrest on federal narcotics charges, much less for the forfeiture of my family's house and land, which often accompanies such an arrest. It was one thing, I reasoned, to grow poppies; quite another to manufacture narcotics from them. I figured I knew where the line between these two deeds fell, and felt confident that I could safely toe it.

But in these days of the American drug war, as it turns out, the border between the sunny country of the law-abiding—my country!—and

*Readers of my last book, *How to Change Your Mind*, as well as the upcoming chapter on mescaline, will perhaps chuckle at this statement.

a shadowy realm of SWAT teams, mandatory minimum sentences, asset forfeitures, and ruined lives is not necessarily where one thinks it is. One may even cross it unawares. As I delved into the horticulture and jurisprudence of the opium poppy last summer, I made the acquaintance of one man, a contemporary and a fellow journalist, who had had his life pretty well wrecked after stepping across that very border. In his case, though, there is reason to believe it was the border that did the moving; he was arrested on charges of possessing the same flowers that countless thousands of Americans are right now growing in their gardens and keeping in vases in their living rooms. What appears to have set him apart was the fact that he had published a book about this flower in which he described a simple method for converting its seedpod into a narcotic—knowledge that the government has shown it will go to great lengths to keep quiet. Just where this leaves me, and this article, is, well, the subject of this article.

1.

Before recounting my own adventures among the poppies, and encounters with the poppy police, I need to tell you a little about this acquaintance, since he was the inspiration for my own experiments in poppy cultivation as well as the direct cause of the first flush of my paranoia. His name is Jim Hogshire. He first came to my attention a few years ago, when this magazine published an excerpt from *Pills-a-go-go*, one of the wittier and more informative of the countless "zines" that sprang up in the early '90s, when desktop publishing first made it possible for individuals single-handedly to publish even the narrowest of special-interest periodicals. Hogshire's own special interest—his passion, really—was the world of pharmaceuticals: the chemistry, regulation, and effects of licit and illicit drugs. Published on multicolored stock more or less whenever Hogshire got around to it, *Pills-a-go-go* printed inside news about the pharmaceutical industry alongside firsthand accounts of Hogshire's own self-administered drug experiments—"pill-hacking," he called it. The zine had a strong libertarian-populist bent, and was given to attacking the FDA, DEA, and AMA with gusto whenever those institutions stood between the American people and their pills—pills that Hogshire regarded with a reverence born of their astounding powers to heal as well as to alter the course of human history and, not incidentally, consciousness.

Hogshire's reports on his drug experiments made for amusing reading. I particularly remember his description, reprinted in this magazine, of the effects of a deliberate overdose of Dextromethorphan Hydrobromide, or DM, a common ingredient in over-the-counter cough syrups and nighttime cold remedies. After drinking eight ounces of Robitussin DM, Hogshire reported waking up at 4:00 a.m. and determining that he should now shave and go to Kinko's to get some copies made.

> That may seem normal, but the fact was that *I had a reptilian brain.* My whole way of thinking and perceiving had changed. . . .
>
> I got in the shower and shaved. While I was shaving I "thought" that for all I knew I was hacking my face to pieces. Since I didn't see any blood or feel any pain I didn't worry about it. Had I looked down and seen that I had grown another limb, I wouldn't have been surprised at all; I would have just used it. . . .
>
> The world became a binary place of dark and light, on and off, safety and danger. . . . I sat at my desk and tried to

> write down how this felt so I could look at it later. I wrote down the word "Cro-Magnon." I was very aware that I was stupid. . . . Luckily there were only a couple of people in Kinko's and one of them was a friend. She confirmed that my pupils were of different sizes. One was out of round . . .
>
> I knew there was no way I could know if I was correctly adhering to social customs. I didn't even know how to modulate my voice. Was I talking too loud? Did I look like a regular person? I understood that I was involved in a big contraption called civilization and that certain things were expected of me, but I could not comprehend what the hell those things might be. . . .
>
> I found being a reptile kind of pleasant. I was content to sit there and monitor my surroundings. I was alert but not anxious. Every now and then I would do a "reality check" to make sure I wasn't masturbating or strangling someone, because of my vague awareness that more was expected of me than just being a reptile. . . .

My interest in Hogshire's drug journalism was mild and strictly literary; as I've mentioned, my own experiments with drugs were past, and never terribly ambitious to begin with. I'd been too terrified ever to try hallucinogens, and my sole experience with opiates had accompanied some unpleasant dental work. I'd grown some marijuana once in the early '80s, when doing so was no big deal, legally speaking. But things are different now: growing a handful of marijuana plants today could cost me my freedom and my house.

We may not hear as much now about the war on drugs as we did in the days of Nancy Reagan, William Bennett, and "Just Say No." But in fact the drug war continues unabated; if anything, the Clinton administration is waging it even more intensely than its predecessors, having spent a record $15 billion on drug enforcement last year and added federal death penalties for so-called drug kingpins—a category defined to include large-scale growers of marijuana. Every autumn, police helicopters equipped with infrared sensors trace regular flight paths over the farm fields in my corner of New England; just the other day they spotted thirty marijuana plants tucked into a cornfield up the road from me, less than a hundred yards, as the crow flies, from my garden. For all I know, the helicopters peered down into my garden on their way; the Supreme Court has recently ruled that such overflights do not constitute an illegal search of one's property, one of a string of recent rulings that have strengthened the government's hand in fighting the drug war.

Overflights and other such measures have certainly proved an effective deterrent with me. And anyway, the few times I've had access to marijuana in the last few years, my biggest problem was always finding the time to smoke it. Whatever else it may be, recreational drug use is a leisure activity, and leisure is something in woefully short supply at this point in my life. No small part of the pleasure I got from reading Hogshire's drug adventures consisted of nostalgia for a time when I could set aside a couple of hours, even a whole day, to see what it might feel like to have a reptilian brain.

Nowadays what leisure time I do have tends to be spent in the garden, a passion that in recent years has turned into a professional interest—I am, among other things, a garden writer. I mention this to help explain the keen interest I took in Jim Hogshire's subsequent project: a somewhat unconventional treatise on gardening titled *Opium for the Masses*, published in 1994 by an outfit in Port Townsend, Washington, called Loompanics Unlimited. The book's astonishing premise is that anyone can obtain opiates cheaply and safely and maybe even legally—or at least beneath the radar of the authorities, who, if

Hogshire was to be believed, were overlooking something rather significant in their pursuit of the war on drugs. According to Hogshire's book, it is possible to grow opium from legally available seeds (he provided detailed horticultural instructions) or, to make matters even easier, to obtain it from poppy seedpods, which happen to be one of the more popular types of dried flowers sold in florist and crafts shops. Whether grown or purchased, fresh or dried, these seedpods contain significant quantities of morphine, codeine, and thebaine, the principal alkaloids found in opium.

Hogshire's claim flew in the face of everything I'd ever heard about opium—that the "right" kind of poppies grow only in faraway places like the Golden Triangle of Southeast Asia, that harvesting opium requires vast cadres of peasant workers armed with special razor blades, and that the extraction of opiates is a painstaking and complicated process. Hogshire made it sound like child's play.

In addition to the horticultural advice, *Opium for the Masses* offered simple recipes for making "poppy tea" from either store-bought or homegrown poppies, and Hogshire reported that a cup of this infusion (which is apparently a traditional home remedy in many cultures) would reliably relieve pain and anxiety and "produce a sense of well-being and relaxation." Bigger doses of the tea would produce euphoria and a "waking sleep" populated by dreams of a terrific vividness. Hogshire cautioned that the tea, like all opiates, was addictive if taken too many days in a row; otherwise, its only notable side effect was constipation.

As for the legal implications, Hogshire was encouragingly vague: "Opium, the juice of the poppy, is a controlled substance but it's unclear how illegal the plant itself is." Here is how I figured one might be able to toe the line safely between the cultivation of opium poppies, routine enough in the gardening world, and felony possession of opium: if opium is the extruded sap of the unripe seedpod, then the dried heads used to make tea *by definition* did not involve one with opium. Hogshire didn't go quite that far, but he did write that "it is unclear whether it is illegal to brew tea from poppies you've purchased legally from the store." As will soon become evident, Jim Hogshire is no longer unclear on either of these points.

Last winter Hogshire's lively little paperback joined the works of Penelope Hobhouse (*On Gardening*), Gertrude Jekyll (*Gardener's Testament*), and Louise Beebe Wilder (*Color in My Garden*) on my bedside table. Winter is when the gardener reads and dreams and draws up schemes for the borders he will plant come spring, and the more I read about what the ancient Sumerians had called "the flower of joy," the more intriguing the prospect of growing poppies in my garden became, aesthetically as well as pharmacologically. From Hogshire I drifted over to the more mainstream garden writers, many of whom wrote extravagantly of opium poppies—of their ephemeral outward beauty (for the blooms last but a day or two) and their dark inward mystery.

"Poppies have cast a spell over gardeners and artists for many centuries," went one typical garden writer's lead; this was, inevitably, quickly followed by the phrase "dark connotations of the opium poppy." But nowhere in my reading did I find a clear statement that planting *Papaver somniferum* would put a gardener on the wrong side of the law. "When grown in a garden," one authority on annuals declared, somewhat ambiguously, "the cultivation of *P. somniferum* is a case of *Honi soit qui mal y pense*. (Shame to him who thinks ill.)" In general the garden writers tended to ignore or gloss over the legal issue and focus instead on the beauty of *somniferum*, which all concurred was exquisite.

Reading about poppies that winter, I wondered if it was possible to untangle the flower's physical beauty from the knowledge of its narcotic properties. It seemed to me that even the lady garden writers who (presumably) would never think of sampling opium had been subconsciously influenced by its mood-altering potential; Louise Beebe Wilder tells us that poppies set her "heart vibrating with their waywardness." Merely to gaze at a poppy was to feel dreamy, to judge by the many American Impressionist paintings of the flower, or from the experience of Dorothy and company, who you'll recall were interrupted on their journey through Oz when they passed out in a field of scarlet poppies. If ever there was an innocent angle from which to gaze at the opium poppy, our culture seems long ago to have forgotten where it is.

By now I too was falling under the spell of the opium poppy. I dug out my college edition of De Quincey's *Confessions of an English Opium-Eater*, and I reread Coleridge's descriptions of his opium dreams (". . . how divine that repose is, what a spot of enchantment, a green spot of fountains and flowers and trees in the very heart of a waste of sands"). I read accounts of the Opium Wars, in which England went to war for no loftier purpose than to keep China's harbors open to opium clipper ships bound from India, whose colonial economy depended on opium exports. I read about nineteenth-century medicine, in whose arsenal opium—usually in the form of a tincture called laudanum—was easily the most important weapon. In part this was because the principal goal of medical care at that time was not so much to cure illness as to relieve pain, and there was (and is) no better painkiller than opium and its derivatives. But opium-based preparations were also used to treat, or prevent, a great variety of ills, including dysentery, malaria, tuberculosis, cough, insomnia, anxiety, and even colic in infants. (Since opium is extremely bitter, nursing mothers would induce babies to ingest it by smearing the medicine on their nipples.) Regarded as "God's own medicine," preparations of opium were as common in the Victorian medicine cabinet as aspirin is in ours.

Is there another flower that has had anywhere near the opium poppy's impact on history and literature? In the nineteenth century, especially, the poppy played as crucial a role in the course of events as petroleum has played in our own century: opium was the basis of national economies, a staple of medicine, an essential item of trade, a spur to the Romantic revolution in poetry, even a *casus belli*.

Yet I had to canvass dozens of friends before I found one who'd actually tried it; opium in its smokable form is apparently all but impossible to obtain today, no doubt because smuggling heroin is so much easier and more lucrative. (One unintended consequence of the war on drugs has been to increase the potency of all illicit drugs: garden-variety marijuana has given way to powerful new strains of sinsemilla; and powdered cocaine, to crack.) The friend who had once smoked opium smiled wistfully as he recalled the long-ago afternoon: "The dreams! The dreams!" was all he would say. When I pressed him for a more detailed account, he referred me to Robert Bulwer-Lytton, the Victorian poet, who'd likened the effect to having one's soul rubbed down with silk.

There was no question that I would have to try to grow it, if only as a historical curiosity. Okay, not *only* that, but that too. Again, you have to understand the gardener's mentality. I once grew Jenny Lind melons, a popular nineteenth-century variety named for the most famous soprano of the time, just to see if I *could* grow them, but also to glean some idea of what the word "melon" might have conjured in the mind of Walt Whitman or Chester Arthur. I planted an heirloom apple tree, "Esopus Spitzenberg," simply because Thomas Jefferson had planted it at Monticello, declaring it the "finest eating apple in the world." Gardening is, among other things, an exercise of the historical imagination, and I was by now eager to stare into the black heart of an opium poppy with my own eyes.

So I began studying the flower sections of the seed catalogs, which by February formed a foot-high pile on my desk. I found "breadseed poppies" (whose seeds are used in baking) for sale in Seeds Blüm, a catalog of heirloom plants from Idaho, and several double varieties (that is, flowers with multiple petals) described as *Papaver paeoniflorum* in the catalog of Thompson & Morgan, the British seed merchants. Burpee carries a breadseed poppy called "Peony Flowered," whose blooms resemble "ruffled pom-poms." In Park's, a large, mid-market seed catalog from South Carolina (their covers invariably feature scrubbed American children arranged in a sea of flowers and vegetables), I found a white double poppy called "White Cloud" and identified as "*Papaver somniferum paeoniflorum.*" Although I didn't know it at the time, all these poppies turn out to be strains of *Papaver somniferum.*

In Cook's, the catalog from which I usually order my seeds for salad greens and exotic vegetables, I found *paeoniflorum* and *rhoeas*, as well as two intriguing varieties of *somniferum*: "Single Danish Flag," a tall poppy that, judging from the catalog copy, closely resembles the classic scarlet poppies I'd read about and seen in Impressionist paintings; and "Hens and Chicks," about which the catalog was particularly enthusiastic: "the large lavender blooms are a wonderful prelude to the seed pods, which are striking in a dried arrangement. A large central pod (the hen) is surrounded by dozens of tiny pods (the chicks)." More to the point, Hogshire had indicated in *Opium for the Masses* that "Hens and Chicks" might prove especially potent.

This was an issue I had wondered about: the ornamental varieties on sale in the catalogs had obviously been bred for their visual or, in the case of the breadseed poppies, culinary qualities. It seemed likely that, as breeders concentrated on these traits to the neglect of others, the morphine and codeine content of these poppies might have dwindled to nothing. So what were the best varieties to plant for opiates?

I couldn't very well pose this question to my usual sources in the gardening world—to Dora Galitzki, the horticulturist who answers the help line at the New York Botanical Garden, or to Shepherd Ogden, the knowledgeable and helpful proprietor of Cook's. So I tried, through a mutual friend, to get in touch with Jim Hogshire himself. I emailed him, explaining what I was up to and asking for recommendations as to the best poppy varieties as well as for advice on cultivation. As I would do with any fellow flower enthusiast, I asked him if he had any seeds he might be willing to share with me and told him about the varieties I'd found in the catalogs. "How can I be confident that these seeds—which have obviously been bred and selected for their ornamental qualities—will 'work'?"

As it turned out, I picked the wrong time to ask. One morning a few days later, and before I'd had any response to my email, I got a call from our mutual friend saying that Hogshire had been arrested in Seattle and was being held in the city jail on felony drug charges. It seems that on March 6 a Seattle Police Department SWAT team had burst into Hogshire's apartment, armed with a search warrant claiming that he was running a "drug lab." Hogshire and his wife, Heidi, were held in handcuffs while the police conducted a six-hour search that yielded a jar of prescription pills, a few firearms, and several bunches of dried poppies wrapped in cellophane. The poppies had evidently come from a florist, but Hogshire was nevertheless charged with "possession of opium poppy, with intent to manufacture and distribute." The guns were legal, but one was cited in the indictment as an "enhancement": another product of the drug war is the fact that the penalties on some narcotics charges rise steeply when the crime "involves" a firearm, even when that firearm is legal or registered. Neither Jim nor Heidi Hogshire had ever been arrested before. Now Jim was being held on $10,000 bail; Heidi, on $2,000. If convicted, Jim faced ten years in prison; Heidi faced a two-year sentence on a lesser charge.

Forgive me for the sudden upwelling of naked self-interest, but all I could think about was that email of mine, buried somewhere on the hard drive of Hogshire's computer, which no doubt was already in the hands of the police forensics unit. Or maybe the message had been intercepted somehow, part of a DEA tap on Hogshire's phone or a surveillance of his email account. I could hardly believe my stupidity! Suddenly I thought I could feel the dull tug of the underworld's undertow, felt as if I'd been somehow *implicated* in something, though exactly what that might be I couldn't say. Yet my confidence that I stood firmly on the sunny side of the law had been shaken. They had my name.

But this was crazy, paranoid thinking, wasn't it? After all, I hadn't *done* anything, except order some flower seeds and write a mildly suggestive piece of email. As for Hogshire, surely there had to be more to this bust than a bunch of dried poppies; it didn't make any sense. I asked our mutual friend if he would be in touch with Hogshire anytime soon, because I was eager to talk to him, to learn more about his peculiar case.

"Also," I added, as casually as I could manage, "would you mind asking him whether he's gotten any email from me?"

2.

My poppy seeds arrived a couple of weeks later. My plan was to sow them, see if I could get flowers and pods, and decide only then whether to proceed any further. I'd been spooked by Hogshire's arrest, doubly spooked to learn from our friend that in fact he had never received my email—undelivered email being highly unusual in my experience. But I still had little reason to doubt that growing poppies for ornamental purposes was legal, and so on an unseasonably warm afternoon in the first week of April I planted my seeds—two packets, each containing a thimbleful of grayish-blue specks. They looked exactly like what they were: poppy seeds, the same ones you find on a kaiser roll or a bagel. (In fact, it is possible to germinate poppy seeds bought from the supermarket's spice aisle. Also, eating such seeds prior to taking a drug test can produce a positive result.)

I'd prepared a tiny section of my garden, an area where the soil is especially loamy and, somewhat more to the point, several old apple trees block the view from the road. *Papaver somniferum* is a hardy annual that grows best in cool conditions, so it isn't necessary to wait for the last frost date to sow; I read that in the South, in fact, gardeners sow their poppies in late fall and winter them over. Sowing is a simple matter of broadcasting, or tossing, the seeds over the surface of the cultivated soil and watering them in; since the seeds are so tiny, there's no need to cover them, but it is a good idea to mix the seeds with a handful of sand in order to spread them as evenly as possible over the planting area.

Within ten days my soil had sprouted a soft grass of slender green blades half an inch high. These were soon followed by the poppies' first set of true leaves, which are succulent and spiky, not unlike those of a loose-leaf lettuce. The color is a pale, vegetal, blue-tinged green, and the foliage is slightly dusted-looking; "glaucous" is the horticultural term for it.

The poppies came up in thick clumps that would clearly need thinning. The problem was, how *much* thinning, and when? Hogshire's book was vague on this point, suggesting a spacing of anywhere from six inches to two feet between plants. My "straight" gardening books advised six to eight inches, but I realized that their recommendations assumed that the gardener's chief interest was flowers. I, of course, was less interested in floriferousness than in, um, big juicy pods. Eventually I called one of the seed companies that sell poppies and delicately asked about optimal spacing, "assuming for the sake of argument someone wanted to maximize the size and quality of his poppy heads." I don't think I aroused any suspicion from the person I talked to, who advised a minimum of eight inches between plants.

Around the time I first thinned my poppies, late in May, a friend who knew of my new horticultural passion sent me a newspaper clipping that briefly stopped me in my tracks. It was a gardening column by C. Z. Guest in the *New York Post* that carried the headline JUST SAY NO TO POPPIES. Guest wrote that although opium poppy seeds are legal to possess and sell, "the live plants (or even dried, dead ones) fall into the same legal category as cocaine and heroin." This seemed very hard to believe, and the fact that the source was a socialite writing in a tabloid not known for its veracity made me inclined to disregard it.

But I guess my confidence had been undermined, because I decided it wouldn't hurt to make sure Guest was wrong. I put in a call to the local barracks of the state police. Without giving my name,

I told the officer who answered the phone that I was a gardener here in town and wanted to double-check that the poppies in my garden were legal.

"Poppies? Not a problem. Poppies have been declared a flower."

I told him the ones I had planted were labeled *somniferum*, and that a neighbor had told me that that meant they were opium poppies.

"What color are they? Are they orange?" This didn't seem especially relevant; I'd read that opium poppies could be white, purple, scarlet, lavender, and black, as well as a reddish-orange. I told him that mine were both lavender and red.

"Those are not illegal. I've got the orange ones in my garden. About two feet tall, came with the house. What you've got to understand is that all poppies have some opium in them. It's only a problem if you start to manufacture opium."

"Like if I slit open a head?"

"Nah, you can cut one of them open and look inside. It's only if you do it with intent to sell or profit."

"But what if I had a *lot* of them?"

"Say you planted two acres of poppies—just for scenery looks? It's not a problem—until you start manufacturing."

I was happy to have the state trooper's okay, but by now a seed of doubt had been planted in my mind. Whether it was C. Z. Guest or the waylaid email—that stupid, incriminating query careening unencrypted through cyberspace—I'd started to get the willies about my poppies. A mild case, to be sure—except for one harrowing night in May when I was caught in the grip of a near-nightmare. In my dream I awake to the sound of police car doors slamming out in front of my house, followed by footsteps on the porch. I leap out of bed and race out the back door into the garden to destroy the evidence. I start eating my poppies, which in the dream are already dried, dry as dust in fact, but I stuff the pods and the stems and the leaves into my mouth as fast as I possibly can. The chewing is horrible, Sisyphean, the swallowing almost impossible; I feel like I am eating my way through a vast desert of plant material, racing madly to beat the clock.

My first impulse on waking was to rip out my poppies right away. My second impulse was to laugh: so this was my first opium dream.

3.

When Jim Hogshire entered my life, in April, my poppies were six inches tall and thriving, their bed a deep, lush carpet of serrated green. I'd heard that Hogshire had raised bail, and our mutual friend was trying to put us in touch; I wanted to talk to him about his case, which I was now thinking of writing about, but I also still hoped to pick up some horticultural tips. I couldn't phone Hogshire, because he'd been thrown out of his apartment. It seems that Washington, like many states, has a law under which tenants charged with drug crimes may be summarily evicted; after the bust, someone from the sheriff's office had paid Hogshire's landlady a visit, notifying her of her "rights" in this regard and urging her to serve the Hogshires with an eviction notice. It sounded to me like a violation of Hogshire's right to due process—after all, he hadn't been found guilty of anything. This was my first introduction to what civil-liberties lawyers have taken to calling "the drugs exception to the Bill of Rights." Over the past several years, in cases involving drugs, the Supreme Court has repeatedly upheld the government's new crop of laws, penalties, and police tactics, thereby narrowing the scope of due process as well as long-established protections against illegal search, double jeopardy, and entrapment.

Hogshire began calling me at odd hours of the day and night. He sounded like a man who had been brought to the end of his tether, edgy and distrustful; disquisitions on *Papaver* nomenclature drifted into diatribes about the indignities his pet birds had suffered at the hands of the police. The voice on the phone was a far cry from the urbane and funny character I'd been reading in *Pills-a-go-go*. But then, Hogshire's bust had left him broke and homeless, bouncing from one friend's couch to another, and adrift on uncharted legal waters—for no one had ever been prosecuted before for possessing dried poppies bought from a florist. Much of what he told me sounded paranoid and crazy, an improbable nightmare featuring a "snitch letter" to the police from a disgruntled houseguest; a search warrant alleging, among other things, that Hogshire was making narcotics out of Sudafed(!); and a police officer who waved Hogshire's writings in his face and asked, "With what you write, weren't you expecting this?" Listening to Hogshire's fantastic account over the phone made me more than a little skeptical, and yet everything he told me I subsequently found confirmed in the court records.

According to documents filed by the prosecutor's office, it was indeed an informant's letter that led to the March 6 raid on the Hogshires' apartment; the letter, sent to the Seattle police by a man named Bob Black, was cited along with Hogshire's published writings as "probable cause" in the search warrant. Bob Black is the disgruntled houseguest, the black hat in Hogshire's bizarre tale. A fellow Loompanics author (*The Abolition of Work and Other Essays*), Black is a self-described anarchist whom the Hogshires met for the first time when he arrived to spend the night on February 10; Loompanics owner Mike Hoy had asked the Hogshires if, as a personal favor, they'd be willing to put Black up in their apartment while he was in Seattle on assignment.

The evening went very badly. Accounts differ on the particulars, as well as on the chemical catalysts involved, but an argument about religion (Hogshire is a Muslim) somehow degenerated into a scuffle in which Black grabbed Heidi Hogshire around the throat and Jim threatened his guest with a loaded M-1 rifle. Ten days later, Black wrote to the Seattle police narcotics unit "to inform you of a drug laboratory . . . in the apartment of Jim Hogshire and Heidi Faust Hogshire." The letter, a denunciation worthy of a sansculotte, deserves to be quoted at length.

> The Hogshires are addicted to opium, which they consume as a tea and by smoking. In a few hours on February 10/11 I saw Jim Hogshire drink several quarts of the tea, and his wife smaller amounts. He also took Dexedrine and Ritalin several times. They have a vacuum pump and other drug-manufacturing tech. Hogshire told me he was working out a way to manufacture heroin from Sudafed.
>
> Hogshire is the author of the book Opium for the Masses which explains how to grow opium and how to produce it from the fresh plant or from seeds obtainable from artist-supply stores. His own consumption is so huge that he must be growing it somewhere. I enclose a copy of parts of his book. He also publishes a magazine Pills a Go Go under an alias promoting the fraudulent acquisition and recreational consumption of controlled drugs.
>
> Should you ever pay the Hogshires a visit, you should know that they keep an M-1 rifle leaning against the wall near the computer.

Largely on the strength of this letter, the police were able to get a magistrate to sign a search warrant and raid the Hogshires' apartment.

It was a quarter to seven in the evening, and Jim Hogshire was reading a book in his living room when he heard the knock at the door; the instant he answered it he found himself thrown up against a wall. Heidi, who was at the grocery store at the time, arrived home to find her husband in handcuffs and a SWAT team, outfitted in black ninja suits, ransacking her apartment. The SWAT team was so large—twenty officers, by Jim's estimate—that only a few could fit into the one-bedroom apartment at a time; the rest lined up in the hall outside.

"Do you publish this?" Jim recalls one officer demanding to know, as he waved a copy of *Pills-a-go-go* in his face. And then, "Where's your poppy patch?" Jim pointed out that it was wintertime and asked the officer, "Why should I grow poppies when they're on sale in the stores?"

"You're lying."

This particular SWAT team specialized in raiding drug labs, which may have been what they expected to find in the Hogshires' apartment. They had to settle, however, for dried poppies: a sealed cardboard box containing ten bunches wrapped in cellophane. The police refused to believe that Hogshire had bought them from a store. The police also found the vacuum pump Black had mentioned (though they didn't bother to seize it), the jar of pills, two rifles and three pistols (all legal), a thermite flare that Hogshire had bought at a gun show, a box of test tubes, and several copies of *Opium for the Masses*.

The Hogshires spent three harrowing days in jail before learning of the charges filed against them. Heidi was charged with possession of a Schedule II controlled substance: the opium poppies. Jim was charged with "possession of opium poppy, with intent to manufacture or distribute," an offense that, with the firearms enhancement, carries a ten-year sentence.

At a preliminary hearing in April, Jim Hogshire was fortunate enough to come before a judge who raised a skeptical eyebrow at the charges filed against him. The hearing had its comic moments. In support of the government's assertion that Hogshire had intent to distribute, the prosecutor, apparently unfamiliar with the literary reference, cited the title of his book: "It's not called 'Opium for Me,' 'Opium for My Friends,' or 'Opium for Anyone I Know.' It's called 'Opium for the Masses.' Which indicates that it's opium for a lot of people."

The judge, a man who evidently knew a thing or two about gardening, found the language in the indictment particularly dubious: the state had accused Hogshire not of manufacturing opium but of manufacturing opium poppies. "How do you manufacture an opium poppy?" the judge asked, and then answered his own question: "You propagate them—it's the only way." By "propagate" the judge meant planting and growing, yet, as he pointed out, the state had presented no evidence that Hogshire had been doing any such thing. "If you had him with a field of poppies, then I think you've got him propagating them in some way. Particularly with the cut poppies and extracting the chemical." But without evidence that Hogshire had actually grown the poppies, the judge reasoned, there was no basis for the manufacturing charge.

The prosecutor sought to recover by citing snapshots seized in the raid that showed Hogshire in an unidentified garden with live poppies whose heads had been slit; he also claimed that "there are poppies outside of his apartment." (There may have been an element of truth to this: according to Hogshire, his landlady had had opium poppies in her garden—though in early March, at the time of the raid, it would have been too early in the season for them to have come up.)

The judge was unpersuaded: "Can you tell me whether those are the relevant genus and species? My mom has poppies outside of her house." The prosecutor could not satisfy the judge on this point, so the judge granted the defense's motion to dismiss the sole charge against Hogshire.

One might think that this would have been the end of Jim Hogshire's ordeal. But the state evidently wasn't through with him, for in June, after dropping charges against Heidi in exchange for a statement asserting that everything seized in the raid belonged to her husband, the prosecutor refiled charges—this time for simple possession of opium poppies—and also added a new felony count to the amended indictment: possession of an "explosive device," citing the thermite flare found during the raid. An arraignment on the new charges was scheduled for June 28. When Hogshire failed to appear, a warrant was issued for his arrest.

4.

I read through the court papers with a mounting sense of personal panic, for the squabble in the Seattle courtroom did not in any way seem to challenge the underlying fact that growing or possessing opium poppies was apparently grounds for prosecution. I called Hogshire's attorney, who confirmed as much and directed me to the text of the Federal Controlled Substances Act of 1970.

The language of the statute was distressingly clear. Not only opium but "opium poppy and poppy straw" are defined as Schedule II controlled substances, right alongside PCP and cocaine. The prohibited poppy is defined as a "plant of the species *Papaver somniferum L.*, except the seed thereof," and poppy straw is defined as "all parts, except the seeds, of the opium poppy, after mowing." In other words, dried poppies.

Section 841 of the act reads, "[I]t shall be unlawful for any person knowingly or intentionally . . . to manufacture, distribute, or dispense, or possess with intent to manufacture, distribute, or dispense" opium poppies. The definition of "manufacturing" includes propagating—i.e., growing. Three things struck me as noteworthy about the language of the statute. The first was that it goes out of its way to state that opium poppy *seeds* are, in fact, legal, presumably because of their legitimate culinary uses. There seems to be a chicken-and-egg paradox here, however, in which illegal poppy plants produce legal poppy seeds from which grow illegal poppy plants.

The second thing that struck me about the statute's language was the fact that, in order for growing opium poppies to be a crime, it must be done "knowingly or intentionally." Opium poppies are commonly sold under more than one botanical name, only one of which—*Papaver somniferum*—is specifically mentioned in the law, so it is entirely possible that a gardener could be growing opium poppies without knowing it. There would therefore appear to be an "innocent gardener" defense. Not that it would do *me* any good: at least some of the poppies I'd planted had been clearly labeled *Papaver somniferum*, a fact that I have—perhaps foolishly—confessed in these very pages to knowing. The third thing that struck me was the most stunning of all: the penalty for knowingly growing *Papaver somniferum* is a prison term of five to twenty years and a maximum fine of $1 million.

So C. Z. Guest had been right after all, and Martha Stewart (and the state trooper) wrong: the cultivation of opium poppies, regardless of the purpose, is indeed a felony, no different in the eyes of the law than manufacturing angel dust or crack cocaine. It didn't matter one bit whether I slit the heads or otherwise harvested my poppies: I had already crossed the line I thought I could safely toe—had crossed it, in fact, back on that April afternoon when I planted my seeds. (What's more, I was vulnerable to the very charge that hadn't stuck to Hogshire—manufacturing!) I was, potentially at least, in deep, deep trouble.

Or was I? For had anyone besides Jim Hogshire ever actually been arrested for the possession or manufacture of poppies? A Nexis search turned up no other case; nor did calls to more than a dozen lawyers, prosecutors, civil libertarians, and journalists who keep tabs on the drug war. Several were unaware that cultivating poppies was even against the law; when so informed, nearly all had precisely the same slightly bemused reaction: "Don't you think the government has better things to do?" I certainly hoped that this was the case, but there the menacing statute was, right there on the books.

I called several experienced gardeners too, hoping to get a clearer picture of the risk involved in growing poppies. One told me a story about a DEA agent on vacation in Idaho who'd tipped off the county sheriff that poppies were being grown in local gardens; another had heard that the DEA had recently ordered the removal of the poppies growing at Jefferson's Monticello. (Both stories sounded apocryphal, but both turned out to be true.) I phoned a radio call-in gardening show, asking the resident expert whether I needed to worry about the opium poppies growing in my garden; "I'm not a lawyer," she said, "but wouldn't it be a shame if gardeners had to pass up such a magnificent flower?"

No one had heard of an actual bust, and most of the gardeners I spoke to seemed blithely unconcerned when I apprised them of the theoretical peril. Some treated me carefully, as though it were paranoid of me to worry. The answer-lady at the New York Botanical Garden tried to reassure me (a bit patronizingly, I thought) by saying that, to her knowledge, there were no "poppy patrols out there." Wayne Winterrowd, the expert on annuals who'd written "Shame to him who thinks ill" of the poppy grower, likened the crime to tearing the tags off pillows and mattresses, another federal offense no one ever seemed to do time for. Laughing off my worries, he offered to send me seeds of a "stunning" jet-black opium poppy he grew in his Vermont garden. He also confirmed (as did a botanist I spoke to later) that "breadseed poppies" as well as *Papaver paeoniflorum* and *giganteum* were botanically no different than *Papaver somniferum*. I'd planted a handful of *paeoniflorum*, and had had no idea what they were—until now.

I took no small comfort in Winterrowd's mattress-tag analogy, if only because I really did not want to have to rip out my poppies, at least not now. For my first poppy was on the verge of bloom. It was the first week of July when I noticed at the end of one slender, downward-nodding stem a bud the size of a cherry, covered in a soft, hairy down. The bud's outer covering, or calyx, had split open, and I could see the scarlet petals folded inside, packed as tightly as a parachute. By the following morning the stem had drawn itself up to its full four-foot height and the petals—five deltas of rich red silk freaked with black—had completely unfurled, casting off their calyx and turning to face the sun. That solitary exquisite bloom was followed the next day by three more equally formidable dabs of pigment, then six, then a dozen, until my poppy patch was a terrific, traffic-stopping blur of color, of a red so red as to be platonic. Now I knew what Robert Browning meant when he spoke of "the poppy's red effrontery": this hue was a shout. The lavender blooms of another variety followed a few days later, a cooler but no less pure jolt of color. When the sun stood behind them, toward evening, the petals were as luminous as stained glass.

"It is a pity," Louise Beebe Wilder wrote, "that Poppies are in such haste to shed their silken petals and display their crowned seedpods." Having seen them, I would have to disagree with her, and not only on pharmacological grounds. The poppy's seedpods are scarcely less arresting than its flowers: swelling blue-green finials poised atop neat round pedestals (called stipes), each pod crowned with an upturned anther like a Catherine wheel. For most of the month of July my whole poppy patch was alive with interest. All at once and side by side you had the drooping sleepy buds, the brilliant flags of color, and the stately upright urns of seeds, all set against the same cool backdrop of dusty green foliage. I couldn't decide what was more beautiful: leaf, bud, flower, or seedpod. I did decide that this poppy patch was as gorgeous as anything I'd ever planted.

My fellow gardeners were making me feel foolish for even thinking of cutting down these flowers; indeed, as I admired my poppies in their full midsummer glory, this unexpectedly lavish gift of nature, it was difficult to credit the notion that they could possibly be illegal—that for the purposes of the law I might just as well be admiring packets of white powder on a table in some dingy uptown drug factory. But this, I knew, was indeed the case. And what a metamorphosis this was!—that an act as ordinary and blameless as the planting of a handful of common and perfectly legal seeds could somehow transport one into the country of criminality.

Yet this was a metamorphosis that required not only the physical seed and water and sunlight but, crucially, a certain metaphysical ingredient too: the knowledge that the poppies I beheld were, in fact, of the genus *Papaver* and the species *somniferum*. For although ignorance of the law is never a defense, in the case of poppies, ignorance of botany may be. True, I had planted seeds I knew to be *Papaver somniferum* and then blabbed that fact to the world. But what if instead I had planted "breadseed poppies," or the poppy seeds on a poppy-seed bagel? What if I had planted only the *Papaver paeoniflorum* I'd ordered, the one I'd had no idea was really *somniferum*? As I stood there admiring the extravagantly doubled blooms of this poppy, I realized that growing it was no more felonious than growing asters or marigolds—for as long, that is, as I remained ignorant of the fact that this poppy, too, was *somniferum*. But it's too late for me now; I know too much. And so, dear reader, do you.

It was precisely this knowledge that inspired the slightly cracked logic behind what I now decided to do. I had not planned to slit even one of my poppies, for fear that it was the step that would take me across the line into criminality. But now I knew I had already taken the fateful step. *In for a dime, in for a dollar.* I know, this wasn't even a remotely rational approach to the situation: a slit seedpod in my garden would constitute proof that I knew exactly what kind of poppies I had. Yet that particular summer afternoon, as I stood there alone with my ravishing poppies, in what, after all, was *my* garden, this logic seemed strangely compelling. So I combed my little stand of poppies for the fattest, most turgid seed head and bent it toward me. Taking the warm, plum-size pod between my thumb and forefinger, I nicked its skin with a thumbnail. After a moment a small bead of milky sap formed on the surface; the wound continued to bleed for a minute or two, the sap darkening perceptibly as it oxidized, and then it slowed, clotting. I dabbed the drop of opium with my forefinger, touched it to my tongue.

It was indescribably bitter. The taste lingered on my palate for the rest of the afternoon.

5.

When I finally met Jim Hogshire in mid-July, it had been two weeks since his missed court date. He was staying in Manhattan, a good place to be anonymous, as he mulled over his next move.

On a hot summer morning we met for coffee on West Twenty-third Street; afterward, we planned to visit the flower district, to shop for dried poppies and check out a rumor that Hogshire had heard about a crackdown on imports of dried poppies. Hogshire was dressed all in white, a slender thirty-eight-year-old with long blond hair gathered in a neat ponytail. His face was handsome but careworn; his fine, angular features were lined, and his deep-set eyes, which are a striking shade of gray, were ringed with shadows. In conversation I found him alternately expansive and wary, though only rarely did he ask to speak off the record. For someone who had no place to live, who was one traffic stop away from going to jail, Hogshire seemed surprisingly composed—or at least a lot more composed than I would be under the circumstances.

Hogshire is passionate about poppies, and we covered that mutual interest for a while, shuttling from *Papaver* horticulture to jurisprudence, *Papaver* nomenclature to chemistry. I learned about the thirty-eight different alkaloids that have been found in *somniferum*, the "biogenetic pathways" from thebaine to morphine (he lost me here), and the "incredible potential" of the "Bentley compounds" that have been synthesized from *Papaver bracteatum*. He told me that he'd first heard about poppy tea from a friend, a gardener whose Russian grandmother had brewed it as a home remedy. Hogshire started experimenting with poppies that he found growing "literally right outside the door of my apartment.

"The first few times I got it all wrong—I didn't grind the poppies up, and I was indiscriminate, using the leaves and stems as well as the pods. I also tried smoking all the various parts, using myself and my wife as guinea pigs. I proved to myself empirically that the heads are undoubtedly the most potent part of the plant." I realized that Hogshire regarded himself as heir to a great tradition of self-experimentation in Western medicine. Eventually he learned how to make a potent tea from dried poppies, pulverizing a handful of heads in a coffee grinder and then steeping the powder in hot water. I asked him to describe the effects of a cup of poppy tea.

"It's not a knock-you-on-your-ass sort of thing, not like smoking opium. In fact, a lot of people will tell you they forget that they are high. It starts with a tickling feeling in the stomach that then rises up into the shoulders and head—this feeling of just . . . *joy*. You feel optimistic about things; energetic but at the same time relaxed. You'll remain functional: you won't say anything stupid and you'll remember everything that happens. You won't nod out, though you will feel a strong desire to close your eyes. Any pain you have will go away; the tea will also relieve exogenously caused depression. That's why poppy tea is served at funerals in the Middle East. It can make sadness go away."

It's hard to believe that commercially available flowers could produce such effects, and at times the claims in Hogshire's book had reminded me of earlier "household highs"—smoking banana peels, for instance ("they call me mellow yellow," Donovan had purred back in

1967), eating morning-glory seeds (purported to be a hallucinogen), or sipping cocktails made from Coca-Cola and aspirin. Could it be there was some sort of placebo effect at work here? Hogshire showed me a scientific article, from the *Bulletin on Narcotics*, that stated plainly that commercially sold dried poppies did indeed contain opiates, in significant quantities. He also pointed out that it was possible to become addicted to poppy tea. In his book he says, "Opium withdrawal hurts, but the pain will end, usually within three to five days. . . . Those are indeed hard days for the kicking addict but it is no worse than a nasty case of the flu." This certainly didn't sound like the effects of a placebo.

If Hogshire was right, then opium was hidden in plain sight in America—which certainly would explain why the government would take an interest in the author of *Opium for the Masses*. He and his small-press book had punctured a set of myths that have served the government well since 1942, when Congress decided that the best way to control opiates was to ban domestic cultivation of *Papaver somniferum* and force pharmaceutical companies to import opium (which they use to produce morphine and other opiates) from a handful of designated Asian countries. Since then the perception has taken hold that this legislative stricture is actually a botanical one—that opium will grow *only* in these places. The other myth Hogshire had exploded is that the only way to extract opiates from opium poppies is by slitting their heads in the field, a complex and time-consuming process that, I heard over and over again from law-enforcement officials and gardeners alike, made the domestic production of opium impractical.

The durability of these myths has obliterated knowledge about opium that was common as recently as a century ago, when opium was still a popular nonprescription remedy and opium poppies an important domestic crop. As late as 1915, pamphlets issued by the U.S.

Department of Agriculture were still mentioning opium poppies as a good cash crop for northern farmers. A few decades before, the Shakers were growing opium commercially in upstate New York. Well into this century, Russian, Greek, and Arab immigrants in America have used poppy-head tea as a mild sedative and a remedy for headaches, muscle pain, cough, and diarrhea. During the Civil War, gardeners in the South were encouraged to plant opium for the war effort, in order to ensure a supply of painkillers for the Confederate Army. The descendants of these poppies are thriving to this day in southern gardens, but not the knowledge of their provenance or powers.

What Hogshire has done is to excavate this vernacular knowledge and then publish it to the world—in how-to form, with recipes. As far as I can tell, the knowledge in his book hasn't seeped too far into the drug culture—*Opium for the Masses* has sold between eight and ten thousand copies, and I turned up no evidence of widespread tea-brewing in drug circles—yet I was curious to know just how far knowledge *about* his knowledge had spread in law-enforcement circles. As Hogshire and I strolled the few blocks up Sixth Avenue to the flower district, he told me that, since the book's publication in 1994, the price of dried poppies had doubled and the DEA had launched a "quiet" investigation into the domestic poppy trade. Agents had paid visits to dried-flower vendors, as well as to the American Association for the Dried and Preserved Floral Industry, a trade group based in Westport, Connecticut. All this sounded to me like either boastfulness or paranoia—until, that is, we got to the flower district.

Manhattan's flower district is modest, a picturesque couple of blocks of lower Sixth Avenue where a few dozen dried- and cut-flower wholesalers have their showrooms at street level. As a pedestrian reaches Twenty-seventh Street, what had been a particularly dreary stretch of

Manhattan suddenly erupts into greenery and bloom. Buckets of dried lotus heads and hydrangeas line the storefronts, gardenias in hanging baskets perfume the air, and clusters of potted ficus trees briefly transform the grubby sidewalk into a fair copy of a garden path. On Twenty-eighth Street we stopped in a narrow, cluttered shop that specializes in dried flowers. Hogshire surveyed a long wall of cubbies stuffed with unlabeled bunches of dried flowers—yarrow, lotus, hydrangeas, peonies, and roses in a dozen different hues—until he spotted the poppies: four different grades, their seedpods ranging in size from marbles to tennis balls, most of them in bunches of ten wrapped in cellophane. The smallest ones still wore a green tint and had a few crunchy leaves wrapped around their stems. The larger poppy heads were buff-colored and strikingly sculptural. They reminded me of a botanical photograph by Karl Blossfeldt, the early-twentieth-century German photographer whose portraits of stems and buds and flowers make them look as if they'd been cast in iron. Hogshire asked the woman at the register if she'd had any problems lately obtaining poppies. She shrugged.

"No problems. How many you need?" I took a bunch, for $10. I felt weirdly self-conscious about my purchase, and the plastic sack she offered me was too short for the long stems, so before we stepped back out onto the street, I turned the bunch head-down in the bag.

We heard a very different story across the street, at Bill's Flowers. Bill told us that he couldn't get poppies anymore; according to his supplier, the DEA—or the USDA, he wasn't sure—had banned imports a few months before, "because kids were smoking the seeds or something." The supplier had told him that it was okay to sell whatever inventory he had left but that there'd be no more poppies after that. Bill's story was my first indication that the federal authorities were, as Hogshire had claimed, doing *something* about the poppy trade—though it would take me several more weeks to figure out exactly what that something was.

Before the morning was over, Hogshire invited me up to his room; the day was getting hot, and he wanted to change his shirt. Most nights since his eviction he'd spent in the apartments of friends, far from home. Tomorrow he expected to be staying somewhere else. I'd asked him earlier why he hadn't stayed to face the charges in Seattle.

"I would go back in a second if I thought they were going to fight fair—if I could be sure they wouldn't manufacture evidence or slap me back in jail at my arraignment. But the fact that they wouldn't just drop this thing after the first charge was thrown out shows me they're being vindictive." (By February, Hogshire had had a change of heart. He said that he'd retained a new lawyer and that he was planning to go back to Seattle to face the charges against him.)

I sat on the bed while Hogshire changed his shirt. Looking around the cramped room, I could see he was traveling light, with little more than a change of clothes, his laptop computer, some books, a stack of articles about poppies, and a sheaf of legal papers about his case. I wondered what it would be like to slip underground—not to be able to go home, not to have your stuff around, not even to know exactly where you would be spending the next night, week, month.

6.

Easy as it may have been to distance myself from Hogshire's underground existence, riding home on the commuter train I found myself wondering just how much circumstantial distance really stood between

Jim Hogshire and me. It was less than meets the eye, and far too little for comfort. I had poppies growing in my garden, after all, and I was preparing an article that would not only acknowledge that fact but would also reprise the very information that had gotten Hogshire into so much hot water. *With what you publish*, the officer had asked Hogshire as they hauled him off to jail, *weren't you expecting this?* So what, exactly, set us apart? For one thing, my life wasn't lived as close to society's margins as Jim's appeared to be; for another, I was writing for a national magazine rather than the fringe press. And this: I didn't associate with people like Bob Black.

I clung to these distinctions in the weeks that followed as I made a concerted effort to learn just how strongly the DEA really felt about poppies—whether, as Hogshire had suggested, the government had launched an investigation and crackdown on domestic opium growing. My curiosity on this point was journalistic but also somewhat more self-interested, and urgent, than that. For by discovering what the DEA was up to, I hoped to learn whether the paranoid fantasies gnawing at me had any basis in reality. I needed to know whether I should be getting rid of my poppies as quickly as possible or whether I could safely let them ripen and then perhaps experiment with poppy tea.

I started checking out Hogshire's leads. At the American Association for the Dried and Preserved Floral Industry, Beth Sherman confirmed that a DEA agent by the name of Larry Snyder had indeed paid the group a visit in 1995. "He asked us to put an article in our newsletter advising people not to carry this certain kind of poppy," she told me. The poppy had always been illegal, the agent had explained to them, but "prior to this they didn't enforce it. They were trying to correct something that had gotten out of hand, but they were trying to do it in a low-key way." The association agreed to publish an article supplied by the DEA informing their membership that it was illegal to possess or sell *Papaver somniferum*.

Hogshire had told me that a Seattle-area flower shop called Nature's Arts, Inc., had also been contacted by the DEA. I got in touch with Don Jackson, the shop's owner. Jackson, who has been in the dried-flower business for forty-five years, told me that a local DEA agent named Joel Wong had visited his shop in March of 1993. The agent had told Jackson that he was investigating poppies and wanted to know what kind his store carried and where they came from.

"He took away several poppies and had them tested. A few weeks later he told me that they were of the opium type and that someone could get high on it, but he didn't say I had to stop selling them." Since then, Jackson had heard rumors of a crackdown and said that he knew of several big domestic growers who had stopped planting poppies for fear of having their crops confiscated. Jackson was concerned about the disappearance of *somniferum* from the trade: "We don't have anything to replace it with," he explained. "That seedpod is so nice and big and round. It's just what people are looking for as a focal point in an arrangement."

When I tried to get in touch with Joel Wong, I learned that he'd recently retired. Another agent in his office took my call but insisted, at the *end* of a fifteen-minute chat, that I not quote him by name. Under the circumstances, I think I'll oblige. Agent Anonymous seemed to be unaware of his predecessor's investigation into dried poppies, so I changed the subject to poppy growing.

"It's illegal to grow opium poppies," the agent said, "but frankly I don't see it becoming a big problem, only because it's so labor-intensive to harvest the opium. You've got to go out early in the morning and slit the pods, then wait until the gum oozes out, and then you have to scrape it off pod by pod. Why would you do all this when you can go down to First and Pike and score some black tar?" (Black tar is a cheap form of heroin from Mexico.) "I say, let 'em at it—it's not going to be a big problem."

It was a friendly enough chat, so I figured I'd ask the agent what advice he'd give a gardener of my acquaintance who had opium poppies growing in his garden. "I'd tell him it's illegal and he's running a risk of getting his front door kicked. But I've got priorities. If he's a University of Washington botanist who's growing poppies, he's not going to have his door kicked; on the other hand, if this professor's scoring the pods, his door most likely will be kicked. It's on a case-by-case basis.

"But I would also tell him, Why grow this illegal plant when there are so many other beautiful plants you can grow? That would be my advice: Why grow the opium when you can put your energy into bonsai plants or orchids, which are so much more challenging? Because how many people can grow an orchid?"

I had told him that I was a garden writer, and he seemed eager to talk about orchid growing, his hobby; he mentioned he kept an orchid on his desk. But when I pressed him about my hypothetical opium-poppy grower, he turned distinctly less amiable.

"What if this poppy grower is also publishing articles about how to make poppy tea?"

"Then his door is going to be kicked. Because he's trying to promote something that's illegal."

It was a chilling conversation. I was reminded of something Hogshire had said about the laws governing opium poppies. "It's as if they had on the books a twenty-miles-per-hour speed limit that was never posted, never enforced, never even talked about. There's no way for you to know that this is the law. Then they pick someone out and say, Hey, you were going fifty. Don't you know the speed limit is twenty? You broke the law—you're going to jail! But nobody else is being stopped, you say. That doesn't matter—this is the law and we have the discretion. The fact that your car is covered with political bumper stickers that we don't like has nothing to do with it. This isn't about free speech!" Whatever else they may be, the drug laws are a powerful weapon in the hands of an Agent Anonymous or, for that matter, a Bob Black. With the speed limit set so low, all it takes is an angry government agent or a "citizen informant" to get you pulled over—to get your door kicked.

It was soon after my conversation with Agent Anonymous that I had my second opium dream. July was nearly over, and I'd come down with a case of Lyme disease, so my nights were already frightful enough, a roller coaster of fevers and bone-rattling chills. In the dream I awake to find faces pressed against the windows of my bedroom, five panes filled with five round white heads: slightly elfin, slightly Slavic-looking. It's a raid, I realize; they're looking for poppies. All night long they search my house, and then, at daybreak, they begin to scour my vegetable garden. They're examining every inch of soil, they're even dusting the leaves of my cabbages for fingerprints. My tormentors are peculiarly non-menacing, and in this dream I've already pulled out my poppies, so I should have nothing to worry about. Even so, I'm trying as hard as I can to watch all five of them at once, just to make sure they don't "plant" anything, but no matter which way I move, one of them is always blocking my view of the others. I move this way, then that, and the frustration of not being able to see what they're up to builds until I think I'm going to explode. And then all of a sudden I spot a single, gorgeous lavender poppy in full bloom on the other side of the garden fence: an escapee. Will they notice it? I wake before I find out, the bedclothes drenched with perspiration.

Maybe the Lyme disease explains the nightmare—I'd had intense, fevered dreams all that week—but it could also have been the call I received from Jim Hogshire earlier that day, announcing that he was thinking of coming up to my place "to help out with the harvest." By comparison, the dream was a walk in the park, for here was a genuine nightmare: I was sick with a 103-degree fever, my joints so stiff I could scarcely turn my head, and a man who was wanted by the police and had no place to live was proposing to come over to help me harvest a crop that could land me in jail. My mind careened as I considered precisely how terrible an idea this was. Did I really want someone who might well, at some point, come under intense pressure from the police (*all right, Hogshire, who else can you finger?*) to see my garden? And once he had unpacked, how was I ever going to get my houseguest to leave? (*The Cable Guy* was in the movie theaters that week.) This is, I know, terribly unfair to Jim Hogshire, who strikes me as a decent-enough fellow, but I kept thinking about something disturbing that he'd told me: that, after his eviction, he had given some serious thought to turning in his landlady for growing opium poppies. I was also flashing on the figure of Bob Black, the Houseguest from Hell. I rifled my brain for a polite and halfway credible excuse, but this was a summit that social etiquette had not yet scaled. In the end I merely spluttered something pathetic about being too sick to think about having people over right now and needing to check with my wife before extending any invitations.

I also told Hogshire that I wasn't sure whether I was *ever* going to harvest, which was true. I didn't yet have a good enough fix on the DEA's intentions regarding poppies and, therefore, on the risk harvesting might entail. It appeared that the DEA was up to something, but *what*, exactly? I knew I should contact the DEA's Washington, D.C., headquarters, but knowing how opaque its agents can be (and being more than a little nervous about alerting them to my existence and interests while my plants were still in the ground), I decided it might be best first to find out as much as I could about the scope of their domestic poppy campaign.

I called Shepherd Ogden at Cook's, one of the seed companies that sells opium poppies. He'd heard rumors that the DEA had sent letters to seed companies requesting they stop selling *somniferum*, though he hadn't received one himself. Ogden reiterated what I already knew: that the sale of seeds is perfectly legal. Beyond that he was uncertain. He suggested that I check with the Association of Specialty Cut Flower Growers, a trade group in Oberlin, Ohio. As it turned out, the president of the association, a northern California flower grower named Will Fulton, had just drafted a column for the latest issue of the association's newsletter alerting members to the DEA letter, which had been received by "one of our most reputable seed companies." The column quoted the letter's first paragraph:

> It has come to the attention of the United States Department of Justice, Drug Enforcement Administration (DEA), that in certain parts of the United States the opium poppy (*Papaver Somniferum L.*) is being cultivated for culinary and horticultural purposes [the italics are Fulton's]. The cultivation of opium poppy in the United States is illegal, as is the possession of "poppy straw" (all parts of the harvested opium poppy except the seeds). Certain seed companies have been identified as selling opium poppy seeds, some with instruction for cultivation printed on the retail packages.

> Before this situation adds to the drug abuse epidemic, DEA is requesting your assistance in curbing such activity.

Judging by the spirited polemic that followed, Will Fulton is the Tom Paine of the cut-flower world. "Wait a minute!" he wrote. "Where's the *mens rea* [criminal intent] here?" Imagine yourself in the interrogation room, he asked his members: "'So, you admit that you intended to cultivate for culinary or horticultural purposes.'

"Why is it illegal to plant a seed, a gift from nature, when your only intention is to grow it for its physical beauty, yet at the same time it is perfectly legal to purchase an AK-47 when your only intention is gopher control?" True, the Founding Fathers had provided for a specific right to bear arms, but the only reason they'd had nothing to say "about the right to plant seeds [was] . . . because it never would have occurred to them that any state might care to abridge that right. After all, they were writing on hemp paper."

When I reached Fulton at his flower farm in northern California, he identified the recipient of the DEA letter as Thompson & Morgan, a venerable British-owned company with offices in New Jersey. Lisa Crowning, the chief horticulturist at Thompson & Morgan, confirmed having received the letter, which she regarded as "intimidating" and "worrisome." Sent by registered mail in late June, the letter was signed by "Larry Snyder, Chief, International Drug Unit"—the same man who'd paid a visit to the American Association for the Dried and Preserved Floral Industry. Thompson & Morgan hadn't yet made a final decision on the DEA's request, but Crowning hoped the firm would continue to offer opium poppies, which she told me she grows in her own garden. Crowning had telephoned Larry Snyder, hoping that there might be "some halfway measure" that would satisfy the DEA (she mentioned putting a warning in the catalog, or removing growing instructions from the packets) but found him completely inflexible. "We don't want to offend the DEA," she told me, "but we feel we are completely within our rights to sell these seeds."

The full text of Snyder's letter to Thompson & Morgan brought the alarming news that the DEA was indeed arresting poppy growers. It alluded to "a recent DEA drug seizure involving a significant quantity of poppy plants . . . many with scored seed pods . . . [that] revealed a supply of poppy seeds noting the date of the shipment and the name and address of your company as the supplier. You should be aware that supplying these seeds for cultivation purposes may be considered illegal." After that thinly veiled threat, Snyder called for a "voluntary cessation of the sale of *Papaver Somniferum L.*"

By October the horticultural grapevine was abuzz with poppy talk and what sounded to me like rumors of war. From Beth Benjamin at Shepherd's Garden Seeds I learned that the police had seized poppies from a public garden project for the homeless that the firm had backed in Santa Cruz. From Will Fulton I learned about a grower in northern California who had had his crop plowed under by the DEA. From the American Seed Trade Association (ASTA) I learned that the DEA—in the person of Larry Snyder—had formally requested that the group call for a voluntary ban on sales of poppy seeds; the association had complied, a staffer told me, "as a civic-duty type of thing." From Katie Sluder, an importer of dried flowers based in North Carolina, I learned that a container load of poppies that she had ordered from a grower in Holland had been turned back by U.S. Customs.

A crackdown was under way, but it was an oddly muffled crackdown. Rather than stage a few well-publicized raids, the DEA seemed to be pursuing a far more subtle strategy. It was working within the industry (in some cases by intimidating companies engaged in legitimate trade) to stanch supplies of both seeds and dried flowers without making any noise in public, much less publicizing exactly what people might be doing with poppies. The subtle hand behind these efforts apparently belonged to Larry Snyder, and I decided the time had come for me to talk to him. When I spotted his phone number printed in ASTA's newsletter, I felt as though I had stumbled upon the Wizard of Oz's direct line.

After I introduced myself as a garden writer, Snyder agreed to an interview. I began by asking his advice on the poppies growing in my garden. He came right to the point: "My advice is not to grow them. It is a violation of federal law. I would get rid of them." He added that "we're not going into Grandma's garden and taking samples of her poppies" and confirmed that a gardener had to be growing *P. somniferum* with knowledge and intent before the deed became a crime.

Perhaps trying to be helpful, Snyder pointed out that there are 1,200 other species of poppies I could be growing instead, including "*rhoeas* and *giganteum* and a jillion others." *Giganteum?* Wasn't that the one Wayne Winterrowd had said was just a strain of *somniferum*? I asked him to describe it. "It's got an even bigger capsule than *somniferum*. I've got one of them sitting right here on my desk."

Snyder acknowledged that the DEA had done nothing to enforce the laws against poppy growing until recently, after receiving "some information coming in out of the Northwest and California that people were making a tea from dried and fresh poppies."

Was he familiar with a book called *Opium for the Masses*?

After what felt to me like an uncomfortably long pause, he said simply, "We see most of the publications."

I might be mistaken, but it was my impression that Snyder grew suddenly curt with me at this point in our conversation. He refused to say anything more about the seizure mentioned in his letter to the seed companies, on the ground that it was "still an active case." When I wondered on what authority the DEA could stop seed companies from selling legal seeds, he cut me off: "If they sell for cultivation purposes, that is illegal." It was hard to see what other reason a seed company would have for selling seeds.

Then I asked Larry Snyder if he worried that his efforts might alert people to just how easy it is to obtain opiates in this country.

"There's always a risk that as more people become aware, some people will try it. It's kind of like announcing that the bank leaves the vault open at nine o'clock in the morning. Is that going to induce someone to rob the bank? Draw your own conclusions."

7.

The conclusion I drew was that the DEA was indeed trying to implement a quiet crackdown, attempting to shut down supplies of poppies, fresh as well as dried, without calling attention to the fact that, as I had discovered with Jim Hogshire's help, they are commonly available and easily converted into a narcotic. What was in the bank vault that Snyder alluded to was this very knowledge, still shut up behind a high wall of misinformation and myth. The DEA appears to be intent on keeping it there, making sure that domestic opium disappears before the knowledge gets out that it is, in fact, hidden in plain sight.

The government would seem to be walking a torturously narrow path here, attempting to send one message to those who are in the know and a very different one to those who are not. This delicate balancing act was on full display in the seizure that Larry Snyder wouldn't discuss with me. I'm fairly sure that I now know what bust Snyder was talking about—or *not* talking about. On June 11, a few weeks before my own poppies had bloomed, the DEA and local law-enforcement agents in Spalding County, Georgia, raided the garden of Rodney Allan Moore, a thirty-one-year-old unemployed man, and his wife, Cherie. Agents seized 258 poppy plants, many of them with their seed capsules scored; two dozen marijuana seedlings; and several ounces of bagged marijuana. A search of the trailer in which the Moores lived turned up records indicating that the poppy seeds had been ordered from Thompson & Morgan and two other firms, as well as a copy of *Opium for the Masses*. Moore was charged with manufacturing morphine and possession of marijuana. Although he had no prior arrest record, he was (and as of February is still being) held on $100,000 bail.*

It does not appear that Moore's bust was part of any organized crackdown on people who grow poppies; acting on an anonymous tip, agents had come looking for a plantation of marijuana and apparently stumbled upon the poppies. But the way the raid was handled is, I think, indicative of the government's two-pronged strategy with respect to domestic opium. While with one hand the DEA took advantage of the bust to track down and apply pressure to the companies that had (legally)

*Moore was indicted by a grand jury on several counts, including manufacturing morphine and possession of a firearm during the commission of a crime. He pled guilty to reduced charges and received a sentence of ten years, of which he served two and a half years, and was ordered to pay a fine of $57,000.

sold Rodney Allan Moore his poppy seeds, with the other it sought to spread a thick cloud of disinformation about poppies before the public.

AGENTS TO CHECK ON HOW POPPIES ENTERED THE COUNTRY, read the page-one headline in the *Griffin Daily News*, alongside a photo of one of Moore's scored poppy heads. The article made no mention of the well-known seed catalogs found in Moore's trailer, which, of course, proved that his poppies had not "entered" the country at all. Instead it quoted Vincent Morgano, a DEA agent, claiming that the growing of opium poppies in this country was unheard of: "In my 25 years with the agency I have never seen it grown in the United States." Clarence Cox, head of the Griffin-Spalding Narcotics Task Force, assured the press that the confiscated poppies are not the same kind that are commonly grown in American flower gardens; Spalding County Sheriff Richard Cantrell said that each of the 258 seedpods seized in the raid could, if properly harvested and processed, yield up to a kilo of heroin apiece. (Talk about alchemy!) Bill Maloney, also with the DEA, explained to a reporter that extracting narcotics from the pods entailed a very complicated and dangerous procedure: "I don't even think someone with a Ph.D. could do it." He also said that opium poppies were extremely rare in the southeastern United States. "The climate has to be just right," he explained. "The temperatures have to be warm and you have to have the right amount of water."

All these assertions I read in the *Griffin Daily News*, which had taken them on faith. And why not? What reason would government officials have to lie about horticulture? Yet several of these statements I had already disproved in my own garden. I knew for a matter of fact that the poppies in question—*Papaver somniferum*—are indeed the same kind commonly grown in American gardens, and that growing them anywhere in the country is not by any stretch a horticultural challenge. And although I did not yet have direct knowledge that these poppies could be made into a narcotic tea, James Duke, a botanist I contacted at the United States Department of Agriculture, had told me that ordinary, garden-variety opium poppies did contain morphine and codeine, and that these alkaloids could easily and effectively be extracted from fresh or dried seedpods by infusing them in hot water—by making a tea. "So you can see why they might be concerned."

And why they might be inclined to lie. If opium is so easy to grow, and opium tea so easy to make, the best—perhaps the only—way for the government to stop people from growing and making their own is to convince them that it can't be done.

I had every reason to believe that James Duke and Jim Hogshire were right, and to doubt the statements of the government agents in Georgia. But it still seemed to me that, in light of the ever-thickening mist of mis- and disinformation swirling around the subject of poppies, the best way to nail down the last piece of poppy knowledge would be to perform a simple experiment on the flowers in my garden. I understood by now that the laws governing poppy cultivation had already expelled me from the country of the law-abiding, indeed had done so even before I knew it had happened. Since those laws drew no distinction between growing poppies and making poppy tea, there seemed to be no good reason *not* to take the steps needed to satisfy my curiosity.

At this point in the story I need to break in to explain why the pages that follow, recounting my "simple experiment," were cut from the original article, on the advice of counsel, and then lost for twenty-four years.

After I submitted the manuscript to *Harper's Magazine* in the late fall of 1996, and while the editing and fact-checking were under way, I mentioned to my editor that we should probably get a lawyer to read the draft, in view of the fact that the government had clearly taken an interest in the activities I was describing, some of which were potentially illegal. John R. "Rick" MacArthur, the publisher of *Harper's*, agreed, and sent the manuscript to a prominent criminal defense lawyer he happened to know. The lawyer practiced in Bridgeport, Connecticut, a city with a long-standing reputation for corruption, organized crime, and illicit drugs—plenty of work for the criminal bar. On a clear winter afternoon, the attorney and his young associate drove up to our home in Cornwall to brief Judith and me on their legal opinion of the piece. It was a weekday and our four-year-old was at day care. We served the lawyers lunch before moving into the living room to hear their counsel. I can remember thinking just how odd it felt to have two criminal defense lawyers in our house, here on business.

Though the senior lawyer spoke in the preternaturally calm tones of his profession, what he had to say terrified us both. If he was right—and I had no reason to doubt him—I was in far more serious legal jeopardy than I had imagined. Throughout the whole experiment, my worst-case scenario, inspired largely by Jim Hogshire's nightmare, had been the midnight visit from the police—the SWAT team armed with a search warrant, tearing up my house and garden while my family and I looked on helplessly. I had always assumed, though, that the government would need some physical evidence (surely the poppies themselves!) or at least an eyewitness—some sort of independent corroboration of the fact that I had grown poppies—before it could bring charges against me.

But after two decades of the war against drugs, the power of the government to move against its citizens has grown even greater than many of us realize. Apparently, a search warrant was the least of my worries. It is at least conceivable that a federal prosecutor could charge me with manufacturing a Schedule II controlled substance with little more evidence than the contents of the article I proposed to publish, which could be admitted into evidence as a confession. The confession could be corroborated with my seed orders, or the felonious poppies that would come up on their own in my garden the next spring, since my poppies had already spread their seed. The penalty? Up to twenty years in prison and a $1 million fine, depending on the quantity of the drug that I was manufacturing. If no poppies were found on the property, under the federal guidelines the government could estimate the amount that *could be grown* in a garden the size of mine and then charge me for growing that.

The lawyer also shared this even more disturbing fact: under federal asset forfeiture laws amended by Congress in 1984 (and since upheld by the Supreme Court*), the government could seize my house and land and evict us from our home without convicting me of any crime, indeed without so much as charging me with one. He explained that my *house and garden* can be "convicted" of the crime of manufacturing opium regardless of whether I am ever charged, let alone convicted, of that offense. Under the civil forfeiture statute, the standard of proof is much lower than in a criminal prosecution; the government need only demonstrate "a preponderance of the evidence" that my property was involved in a violation of the drug laws in order

*In 2019 the Court put some limits on civil forfeiture, citing the Eighth Amendment's bar against "excessive fines."

to confiscate it. What would it take to establish that preponderance? In the opinion of the lawyer seated across from me in our living room, nothing more than the article I was planning to publish.*

As I listened to this attorney calmly explain how the act of publishing this story could wreck our lives, I could see that there were two narratives at war here. In my version of the story, it would be no big deal to harvest a couple of seedpods from my garden, crush and steep them in a cup of hot water, and taste the resultant tea, which I thought of as a fairly mild herbal remedy. But that's my description. The lawyer was telling me I had to weigh, if not accede to, the government's very different description of those same acts: that making poppy tea is "manufacturing narcotics"; that printing its recipe and describing its effects in any but the most horrific terms would be "promoting drug abuse." The decision whether to prosecute a person turns not only on what crimes he may or may not have committed but also on what kind of story a prosecutor can tell a jury about him, and according to the lawyer, the government's version of the story might well prevail over mine. My situation was made worse by the fact that there was no way to disguise either where or when the crime I would be confessing to in print took place: the events are obviously set in my house and garden (thereby establishing the jurisdiction and target asset for forfeiture), and the exact time the crime took place can easily be ascertained by dating events in the narrative, such as Hogshire's arrest, making it impossible for me to claim the

*You might ask, as I did the lawyer, whether the fact that I am a journalist growing poppies for the purpose of writing about them offered me any protection, under the First Amendment or a state shield law. The answer is no. There was no shield law in the state of Connecticut in 1996, and even if there were, shield laws offer no protection to a journalist engaged in a criminal activity.

statute of limitations had passed. From an evidentiary point of view, my article was a bonfire of self-incrimination.

The decision whether to proceed, or not, was mine, the lawyer said in concluding, but he could not as my attorney advise publication.

I was flabbergasted. Sitting in my own living room, on my familiar sofa, I suddenly felt as though I'd metamorphosed into another kind of being—a defendant, and one whose goose was well and truly cooked. The decision before me seemed obvious: I'd be a fool to jeopardize not only my freedom but our home by publishing an article.

It wasn't just any article, however. I had spent the better part of a year working on it and, as a freelance writer, was counting on the fee. But even before the lawyers packed up their briefcases and headed back to Bridgeport, I could see all that effort and income swirling down the drain of my stupidity. *What had I been thinking?*

But the story didn't end there, obviously, since I did ultimately publish the piece—or at least *most* of it. When word of the lawyer's advice and my reaction to it reached Rick MacArthur, he was outraged. It's important to understand that Rick is not your typical magazine publisher, one with an eye fixed on the bottom line and a genetic aversion to litigation. Rick is fierce in his devotion to press freedom and has a tropism bending him toward, rather than away from, the bright light of controversy. The recommendation of his lawyer friend to suppress a piece of journalism for any reason was an affront to his very being.

Rick's immediate response?

Find a new lawyer!

This time, instead of a criminal defense lawyer, Rick hired a First Amendment lawyer, one of the most prominent in New York

City. Victor Kovner had represented numerous well-known authors, filmmakers, and media outlets, often defending them from government efforts to suppress their work. Victor read the same draft the Bridgeport lawyer had read but came to the opposite conclusion. I don't recall his exact words, but what I heard was: *This piece* must *be published for the good of the republic!* He deemed it unlikely that the government would come after a magazine as well-known and venerable as *Harper's*. In his view the piece should be read not as a confession to a crime but rather as a political commentary on the drug war, the precise type of speech the First Amendment exists to protect. Together Kovner and MacArthur made me feel that my concerns—for my liberty, for my home!—were parochial when set against the public interest at stake. If anything, they seemed eager for a fight.

What to do? I was badly torn. I very much wanted to publish a piece I was proud of and—no small matter—get paid for it. Maybe the Connecticut lawyer was overreacting, and failing to weigh the political calculation that the government would be foolish to go after us. Shouldn't I, as a journalist, look beyond my own safety and give at least some weight to the First Amendment issues hanging in the balance here?

I pressed Rick to see how far he and the magazine would go to defend me in the event something happened. In reply, he had Kovner draft a letter of agreement, which stands as one of the most unusual contracts ever given to any writer by a publisher. If anything happened to me as a result of the publication of the article, *Harper's* committed to "defend, indemnify you and hold you harmless from and against any and all costs, expenses and losses of any kind." This included not only paying for my defense (and promising not to settle a case without my consent), but reimbursing me for the time spent defending myself. In the event that I lost a case and was incarcerated, *Harper's* agreed to pay a salary to Judith until my release, as well as any fines or penalties. And if the government should seize our house and land, *Harper's* committed to buying us a comparable new home. The agreement was reassuring but it was also frightening to read: *All these contingencies could actually happen.*

I asked Kovner whether there was anything I could do to protect myself if, in fact, I was willing to publish. He suggested that there were two passages in the piece that were most likely to antagonize the government, and if I could live without them, it might reduce the likelihood of prosecution. As I recall, he cited *United States v. Progressive Inc.*, a 1979 case in which the government had sought to stop *The Progressive* magazine from publishing an article containing instructions for making a hydrogen bomb, even though the instructions were based entirely on publicly available information.* By publishing a recipe for making poppy tea, and then describing its effects in generally positive terms, I would be seen as taunting the government as well as educating would-be opium growers; in Kovner's judgment this increased the likelihood that the government would feel compelled to take some kind of action. Removing those pages would minimize that risk, he felt, since the article would then, in effect, be serving the DEA's purpose: intimidating people like me from divulging the recipe for poppy tea and describing its effects. Kovner also felt that a defendant who hadn't used the drug in question would be more sympathetic in the eyes of a jury. But his bottom

*The government ultimately dropped the case during appeals, declaring the issue moot after much of the information contained in the piece had become public.

line was that if I was willing to cut the offending pages, I could reduce my exposure to "negligible."

So that is what, after consulting with Judith and agonizing for several days, I decided to do. I cut the recipe and "trip report" and, before the magazine went to print, made sure to get those passages, along with any other potential evidence, off the property and my computer. But before erasing it from my hard drive, I copied the unexpurgated version of the piece to a floppy disk and gave it to my brother-in-law, an attorney, for safekeeping. Why? I couldn't bear to destroy it. Maybe someday, I thought—after the drug war ended or the statute of limitations had passed—I would do something with it.

Here are those missing passages, followed by the final section of the piece as it appeared in 1997.

8.

It was late fall when I finally harvested my poppies. By now they had dried on their stalks, forming crinkled brown seedpods the size of walnuts.

According to James Duke, the retired USDA researcher I had spoken to, I had passed up a pharmacological opportunity by failing to harvest the seedpods while still fresh and full of sap, or opium. Duke suggested that alcohol would make a better solvent than hot water for extracting alkaloids from poppies, which made sense: laudanum is a name for just such a tincture of opium. "You can get the equivalent of a shot of heroin from a good green pod dissolved in a glass of vodka," Duke told me. I wondered why Hogshire's recipes focused on poppy tea to the exclusion of alcohol-based preparations and then recalled something he'd told me: Hogshire was a Muslim, and so didn't drink alcohol.

Examining the pods in my garden, I could see that the tiny portals circling the anther at the top of each capsule had opened, releasing the poppy seeds to the wind. The seed portals looked exactly like the little observation windows circling the crown of the Statue of Liberty. By now the seeds had probably been dispersed all over my garden, and would come up on their own, willy-nilly, next spring. If I didn't want opium poppies next season, I would have to sedulously weed every one of these volunteers.

I snapped a half dozen of the pods off their stalks and brought them into the kitchen. Though many of their seeds had been dispersed, many more remained, and the pods made a rattling sound whenever they moved. Following Hogshire's recipe, I shook out the rest of the seeds (there were hundreds in each pod, ranging in color from beige to lavender to black) and crushed the pods in my fist. The shards I stuffed into the bowl of a coffee grinder, which in a few seconds noisily reduced them to a fine dun powder. I boiled a kettle of water and poured it over the dry tea in a mug, stirred the chestnut-colored mixture, and let it steep. The aroma was not at all unpleasant; it smelled of hay, not unlike a lapsang souchong tea. The whole procedure was so straightforward, so domestic in its particulars, that it felt no more controversial than making pesto or lemon balm tea, two equally simple harvest operations I'd performed that week. I certainly didn't feel the lack of a Ph.D.

After fifteen minutes I poured the tea through a strainer, in the well of which it deposited a viscous brown slurry. With the back of a tablespoon I mashed this material against the mesh of the strainer, pushing through the last few ounces of liquid. The tea was ready to drink.

Poppy tea tastes truly awful. It was nearly as bitter as raw opium and, after the novelty of the flavor wore off, slightly nauseating. I had asked James Duke why he thought poppies produced opium in the first place—what, in other words, was the evolutionary point? Alkaloids taste bad, he pointed out; it's conceivable that plants produce them as a defense against pests. "No animal's going to bother a plant that tastes that bad. So the plant with the worst taste is going to produce the most offspring."

It was a job getting a cup of the stuff down. The tea not only tasted terrible, but it was oddly filling too, and very soon made me queasy, a sensation much like a mild seasickness. I wondered if it was even possible to overdose on poppy tea; it seemed to me your stomach would rebel long before a significant amount could be ingested.

Within ten minutes or so, I began to feel . . . different. Not dramatically different, not "high," but not exactly the same self I was ten minutes before, either. Remembering what Jim Hogshire had told me about the tea's analgesic properties, I conducted an inventory of my everyday aches and pains and physical annoyances—a stiffness in the neck I'd woken with, the nasal and throat irritations of a particularly bad hay fever season, the usual dull pain in my knuckles after too many hours at the computer keyboard—and found that all these symptoms had, if not quite disappeared, then dropped beneath the threshold of my attention. They simply didn't matter. Then I decided it would be a good idea to inventory my mood, and concluded that it was very good indeed. Nothing I would describe as euphoric, but I was suffused body and mind with a distinct feeling of well-being—the words "warm" and "aqueous" appear in my notes. I'm not sure whether it was the mode of self-study I had logged onto, but the mental stance of standing just slightly apart from my self, coolly appraising my sensations and moods, suddenly seemed like the most natural thing in the world. I felt as though I was almost, but not quite, having an experience in the third person.

Hogshire had said that the tea "can make sadness go away," and now I understood why he had employed that particular phrasing. For the poppy tea didn't seem to add anything new to consciousness, in the way that smoking marijuana can produce novel and unexpected sensations and emotions; by comparison, the tea seemed to subtract things: anxiety, melancholy, worry, grief. Like the opiate it is, or consists of, poppy tea is a pain killer in every sense. In my notes I wrote "definitely lightens the existential load."

Fully expecting to be rendered useless by the tea—I have always been highly susceptible to drugs, and opiates are commonly thought to be soporific—I had chosen an afternoon for my experiment on which there was little I needed to get done. And for the first hour, as I sat there at my desk assessing its effects, I did feel a powerful urge to close my eyes—not from any drowsiness, but from a radical and by no means unpleasant sense of passivity. I just didn't need to have all that visual information, thank you very much. My senses were functioning normally, yet I didn't particularly feel like acting on their data. At one point I remember feeling chilled, but couldn't be bothered to close a window or put on a sweater. I'll just sit here awhile longer if that's okay. "Like sitting out on the front porch of one's consciousness, watching the world go by," I wrote, somewhat cryptically.

But I found I could think clearly—as long as I thought about one thing at a time. De Quincey had said he found reading a congenial activity while eating opium, and for a while I read a book with perfect concentration. But during the second hour I noticed I was actually feeling energetic, even purposeful. Now I felt like stepping down from porch-consciousness and heading out into the garden to take care of a few chores.

This was to be, I had decided beforehand, a one-time experiment, and I knew I had to rid my garden of poppies, the sooner the better. So I set to work pulling up the withered stalks. But I was unsure exactly what to do with this crop of dead flowers—this evidence. I had read that the police no longer needed a search warrant to search my garbage (another juridical fruit of the drug war), so throwing them out with the trash was out of the question. I finally decided simply to compost them; by spring they'd be indistinguishable from the decomposing sunflower heads, broccoli plants, eggshells, and table scraps mounded up on the pile of compost in the corner of my vegetable garden.

9.

As I gathered up the poppy stalks, I reflected on the season's unusual harvest. Pride is a common enough emotion among gardeners at this time of year—that, and a continuing amazement at what it is possible to create, virtually out of nothing, in one's garden. I still marvel each summer at the achievement of a Bourbon rose or even a beefsteak tomato—how the gardener can cause nature to yield up something so specifically attractive to the human eye or nose or taste bud. So it was with these astonishing poppies: how can it be that such an inconsequential speck of seed could yield a fruit in my garden with the power to lift pain, alter consciousness, "make sadness go away"?

We have the scientist's explanation: the alkaloids in opium consist of complex molecules nearly identical to the molecules that our brain produces to cope with pain and reward itself with pleasure, though it seems to me that this is one of those scientific explanations that only compounds the mystery it purports to solve. For what are the odds that a molecule produced by a flower out in the world would turn out to hold the precise key required to unlock the physiological mechanism governing the economy of pleasure and pain in my brain? There is something miraculous about such a correspondence between nature and mind, though it too must have an explanation. It might be the result of sheer molecular accident. But it seems more likely that it is the result of a little of that and then a whole lot of co-evolution: one theory holds that *Papaver somniferum* is a flower whose evolution has been directly influenced by the pleasure, and relief from pain, it happened to give a certain primate with a gift for horticulture and experiment. The flowers that gave people the most pleasure were the ones that produced the most offspring. It's not all that different from the case of the Bourbon rose or the beefsteak tomato, two other plants whose evolution has been guided by the hand of human interest.

There was a second astonishment I registered out there that autumn afternoon, this one somewhat darker. As I threw my broken stalks on the compost and turned them under with a pitchfork, I thought about what it could possibly mean to say that this plant was "illegal." I had started out a few months ago with a seed no more felonious than the one for a tomato (indeed, they had arrived in the same envelope), and, after planting and watering it, thinning and weeding and performing all the other ordinary acts of gardening, I had ended up with a flower that rendered its cultivator a criminal. Surely this was an alchemy no less incredible than the one that had transformed that same seed into a chemical compound with the power to alter the ratio of pleasure and pain in my brain. Yet this second transformation had no basis in nature whatsoever. It is, in fact, the result of nothing more than a particular legal taxonomy, a classification of certain substances that appear in nature into categories labeled "licit" and "illicit." Any such taxonomy, being the product of a particular culture and history and politics, is an artificial construct. It's not difficult to imagine how it might have been very different than it is.

In fact it once was, and not so long ago. Not far from my garden stands a very old apple tree, planted early in this century by the farmer who used to live here, a man named Matyas, who bought this land in 1915. (The name is pronounced "matches.") The tree still produces a small crop of apples each fall, but they're not very good to eat. From what I've been able to learn, the farmer grew them for the sole purpose of making hard cider, something most American farmers had done since Colonial times; indeed, until this century hard cider was probably the most popular intoxicant—drug, if you will—in this country. It shouldn't surprise us that one of the symbols of the Women's Christian Temperance Union was an ax; prohibitionists like Carry Nation used to call for the chopping down of apple trees just like the one in my garden, plants that in their eyes held some of the same menace that a marijuana plant, or a poppy flower, holds in the eyes of, say, [drug czar] William Bennett.

Old-timers around here tell me that Joe Matyas used to make the best applejack in town—100 proof, I once heard. No doubt his cider was subject to "abuse," and from 1920 to 1933 its manufacture was a federal crime under the Eighteenth Amendment to the Constitution. During those years the farmer violated a federal law every time he made a barrel of cider. It's worth noting that during the period of anti-alcohol hysteria that led to Prohibition, certain forms of opium were as legal and almost as widely available in this country as alcohol is today. It is said that members of the Women's Christian Temperance Union would relax at the end of a day spent crusading against alcohol with their cherished "women's tonics," preparations whose active ingredient was laudanum—opium. Such was the order of things less than a century ago.

The war on drugs is in truth a war on *some* drugs, their enemy status the result of historical accident, cultural prejudice, and institutional imperative. The taxonomy on behalf of which this war is being fought would be difficult to explain to an extraterrestrial, or even a farmer like Matyas. Is it the quality of addictiveness that renders a substance illicit? Not in the case of tobacco, which I am free to grow in this garden. Curiously, the current campaign against tobacco dwells less on cigarettes' addictiveness than on their threat to our health. So is it toxicity that renders a substance a public menace? Well, my garden is full of plants—datura and euphorbia, castor beans, and even the leaves of my rhubarb—that would sicken and possibly kill me if I ingested them, but the government trusts me to be careful. Is it, then, the prospect of pleasure—of "recreational use"—that puts a substance beyond the pale? Not in the case of alcohol: I can legally produce wine or hard cider or beer from my garden for my personal use (though there are regulations governing its distribution to others). So could it be a drug's "mind-altering" properties that make it evil? Certainly not in the case of Prozac, a drug that, much like opium, mimics chemical compounds manufactured in the brain.

Arbitrary though the war on drugs may be, the battle against the poppy is surely its most eccentric front. The exact same chemical compounds in other hands—those of a pharmaceutical company, say, or a doctor—are treated as the boon to mankind they most surely are. Yet although the medical value of my poppies is widely recognized, my failure to heed what amounts to a set of regulations (that only a pharmaceutical company may handle these flowers; that only a doctor may dispense their extracts) and prejudices (that refined alkaloids are superior to crude ones) governing their production and use makes me not just a scofflaw but a felon.

Someday we may marvel at the power we've invested in these categories, which seems out of all proportion to their artifice. Perhaps one day the government won't care if I want to make a cup of poppy tea for a migraine, no more than it presently cares if I make a cup of valerian tea (a tranquilizer made from the roots of *Valeriana officinalis*) to help me sleep, or even if I want to make a quart of hard apple cider for the express purpose of getting drunk. After all, it wasn't such a long time ago that the fortunes of the apple and the poppy in this country were reversed.

As I made sure the stalks were well interred beneath layers of compost, close enough to the heat at the center of the pile to blast them beyond recognition, I thought about how little had changed in my garden since Joe Matyas tended it during Prohibition, a time we rightly regard as benighted—and wrongly regard as ancient history. If anything, those of us living through the drug war live in even stranger times, when certain plants themselves have been outlawed from our gardens with no regard for what one might or might not be doing with them. Prohibition never outlawed Joe Matyas's apple trees (nor did it threaten this property with confiscation); it wasn't until Matyas made his cider that he crossed the line.

But there it was, then as now, a line through the middle of this garden. Thanks to two national crusades against certain drugs that can be easily produced in it, both he and I found a way to violate federal law without so much as stepping off the property, and jeopardized our personal freedom simply by exercising it. In addition to inhabiting this particular corner of the earth, Matyas and I presumably had a few other things in common. There is, for example, the desire to occasionally alter the textures of consciousness, though I wonder if that might not be universal. And then there's this: the refusal to accept that what happens in our gardens, not to mention in our houses, our bodies, and our minds, is anyone's business but our own. Fifteen years ago, when I first moved into this place, some of the crumbling outbuildings dotting the property still bore crudely lettered warnings directed, I liked to think, at the dreaded "Revenuers" and anyone else the old farmer judged a threat to his privacy—to his liberty. KEEP OUT! went one, an angry scrawl painted in red on the side of a shed. My sentiments exactly.

Epilogue

You're probably wondering what happened after the article was published. I spent a few anxious weeks waiting for some shoe to drop, but either the government never saw the piece (unlikely in view of what happened to Hogshire's obscure book) or Kovner's political calculation was correct, and the government decided it had more to lose by coming after us than it stood to gain. If the crackdown on domestic opium production was intended to be a quiet one, aimed at stopping the activity without alerting anyone to its existence, a noisy battle with a national magazine would surely undermine that strategy. But, of course, all this is speculation: who knows what they were thinking, assuming they paid the matter any attention at all?

And who knows if my act of self-censorship made the difference. I came to regret cutting the pages from the piece, though not until the fear and paranoia that gripped me that year had subsided. It takes no courage to publish the offending pages now; the statute of limitations on my crimes passed years ago. No, the only problem with publishing the missing pages now was finding them.

I thought I had left the pages in the custody of my brother-in-law; however, when I asked after them recently, he claimed to have returned the files to me many years ago. I had no recollection of getting them back. But when I mounted a serious search among my papers, I found—in a storage closet under the daybed in my writing studio in Cornwall—a thick old-school legal folder containing some faxed galleys of the piece, some legal memos, drafts of the *Harper's* indemnification letter, and a single purple floppy—a Zip drive. I was hopeful this might be it—but I had no machine that could read the ancient and obsolete disk.

After asking around, I heard about a computer consultant in a neighboring town named David Maffucci, by reputation a wizard at this sort of thing. When I reached Dave by phone, he said that he had a basement full of "old media," and might have something that could read my disk, provided it hadn't deteriorated too badly. I dropped it off at his shop. Days later, Dave called to report that he had managed to find the right machine, and the contents of the disk were intact and readable. He copied them onto a thumb drive. On it I found a dozen Microsoft Word files related to the piece with one titled, promisingly, "poppy draft 11-1 copy." That had to be it.

But there was a problem: the then-current version of Word couldn't open files from that distant era. Thankfully, Dave once again had the workaround. He pointed me to a free piece of software that I could download from the net called LibreOffice. Miraculously, LibreOffice was able to open the file, and there it was, a first draft complete with the recipe and trip report you've just read, words I hadn't laid eyes on in twenty-four years.

If there is a lesson to this part of the story, it is that the best way to save information for more than a handful of years is not digital technology, but acid-free paper.

Opium, Made Easy," as *Harper's* called the version it published, did not launch a nationwide fad for DIY opium production, as far as I could tell. I did hear anecdotally that sales of *Papaver somniferum* seeds were unusually robust the following year, though it took gardeners some effort to find them in the seed catalogs; several companies had dropped the flower or changed the name that it was sold under after coming under pressure from the DEA.

But whatever the DEA was thinking in 1996 and '97, the government missed the real story about opium, as in fact did I. While we were caught up in this remote and ridiculous skirmish in the drug war, the drug in question was quietly and legally making its way into the bodies of millions of Americans, as Purdue Pharma pursued its marketing campaign, seeding the culture with seductive disinformation about the safety of OxyContin. There's a parable here somewhere, about the difference between journalism and history. What might appear to be "the story" in the present moment may actually be a distraction from it, a shiny object preventing us from seeing the truth of what is really going on beneath the surface of our attention, what will most deeply affect people's lives in time. This also turns out to be a pretty good summary of the drug war, which, besides doing so much to erode our liberties and fill our prisons, served to distract us from reckoning the true toll of the opiates we happened to classify as legal.

I mentioned earlier that you don't hear nearly as much about the drug war anymore. Efforts are afoot to undo some of its damage and decriminalize some of the plants it demonized, though even the Decriminalize Nature movement, which seeks to exempt illicit "plant medicines" from prosecution, won't touch opium—such is the stigma that the opioid crisis has stamped on that flower and its medicine. But though it is now widely recognized that the drug war has been a failure, to judge by the number of arrests for violations of the drug laws, it might as well be 1997: 1,247,713 arrests then; 1,239,909 in 2019. If the drug war is over, the police and the DEA apparently haven't gotten the memo yet.

As for the Sacklers and their criminal enterprise, at least some small portion of justice has been done. In 2020, the family agreed to a settlement with the Department of Justice, under which they pled guilty to criminal charges and agreed to pay $8.3 billion in penalties. Early in 2021, the Sacklers proposed an additional $4.275 billion to reimburse states, municipalities, and tribes for costs incurred by the epidemic and to compensate the families of their victims—the hundreds of thousands of people who have died by opioid overdose since the introduction of OxyContin in 1996. It is unfortunate that, thanks to the protections afforded by the bankruptcy laws and the ingenuity of lawyers and accountants, it could be years before any of these families sees a dime.

And Jim Hogshire? He managed to avoid jail time and got off with a fine, community service, and a year of probation. In the years since, he seems to have fallen on hard times, but whether this owes to his encounter with the drug war, I can't say. He doesn't appear to have published anything since the 1990s. The last mention of him in the press I could find was from 2014, when he was interviewed for an article about people who live in their cars on the streets of Seattle, under the threat of having their "homes" impounded for unpaid parking tickets. Jim and Heidi were living in a camper parked on the street; his battle now was not with the DEA but with the meter maids. He told the reporter: "This is the step you get before you become totally homeless."

CAFFEINE

Maybe the very first sentence isn't the best place to admit this, at the very moment you are deciding whether to grant me an hour or two of your attention, but halfway through the research for this story, I suffered a crisis of confidence that caused me to doubt the subject was of any interest at all, even to me, whose supposedly bright idea it was. I began seriously to doubt a long piece on caffeine was worth the time and effort it would take to report and write it, and to wonder why I had ever thought otherwise. I was in trouble. *We* were in trouble. Though you have an option, I do not: you, at least, can stop reading right here.

Before this crisis, I had been chugging merrily along, conducting interviews, reading countless books of science (it turns out caffeine is one of the most studied psychoactive compounds there is) and history (the course of which was shifted decisively in the West by the introduction of caffeine); traveling to South America to visit a coffee finca; tasting all manner of caffeinated beverages, when suddenly, like Wile E. Coyote in the Road Runner cartoon, I chanced to glance down and realized there was no more road underfoot, just a vast empty expanse of pointlessness as far as I could see. *What in the world was I doing?*

Or perhaps it would be more accurate to ask, What was I *not* doing? Because something was going on with me just then that almost certainly accounts for this project's sudden loss of cabin pressure: I had stopped using caffeine. Abruptly and completely.

After years of a tall morning coffee, followed by several cups of green tea throughout the day, and the occasional cappuccino after lunch, I had quit caffeine, cold turkey. It was not something that I particularly wanted to do, but I had come to the reluctant conclusion that the story demanded it. Several of the experts I was interviewing had suggested that I really couldn't understand the role of caffeine in my life—its invisible yet pervasive power—without getting off it and then, presumably, getting back on. Roland Griffiths, one of the world's leading researchers of mood-altering drugs, and the man most responsible for getting the diagnosis of "Caffeine Withdrawal" included in *The Diagnostic and Statistical Manual of Mental Disorders* (or the *DSM-5* for short), the bible of psychiatric diagnoses, told me he hadn't begun to understand his own relationship to caffeine until he stopped using it and conducted a series of self-experiments. He urged me to do the same.

The idea here is that you can't possibly describe the vehicle you're driving without first stopping, getting out, and taking a good look at the thing from the outside. This is probably the case with all psychoactive drugs, but is especially true of caffeine, since the particular quality of consciousness it sponsors in the regular user feels not so much altered or distorted as normal and transparent. Indeed, for most of us, to be caffeinated to one degree or another has simply become baseline human consciousness. Something like 90 percent of humans ingest caffeine regularly, making it the most widely used psychoactive drug in the world, and the only one we routinely give to children (commonly in the form of soda). Few of us even think of it as a drug, much less our daily use of it as an addiction. It's so pervasive that it's easy to overlook the fact that to be caffeinated is not baseline consciousness but, in fact, an altered state. It just happens to be a state that virtually all of us share, rendering it invisible.

So I decided that for the good of the piece—that is to say, for *you*, dear reader—I would conduct a self-experiment in abstention. What had never occurred to me when I began this experiment is that, by giving up caffeine, I would be undermining my ability to tell the story of caffeine, a knot I wasn't at all sure how to untie.

Maybe I should have anticipated the problem. The scientists have spelled out, and I had duly noted, the predictable symptoms of caffeine withdrawal: headache, fatigue, lethargy, difficulty concentrating, decreased motivation, irritability, intense distress, loss of confidence(!), and dysphoria—the polar opposite of euphoria. I had them all, to one degree or another, but beneath the deceptively mild rubric of "difficulty concentrating" hides nothing short of an existential threat to the work of the writer. How can you possibly expect to write anything when you can't concentrate? That's pretty much all writers do: take the blooming multiplicity of the world and our experience of it, *literally* concentrate it down to manageable proportions, and then force it through the eye of a grammatical needle one word at a time. It's a miracle anyone ever manages this mental feat, or at least it seems that way on day three of caffeine withdrawal. But even before the writer can hope to confront and scale that sheer cliff of impossibility, he or she needs to muster the confidence—the sense of agency and power—required to proceed. It hardly matters if it's a delusion, but that sense that you have by the tail a story the world needs to hear, and you alone have what it takes to tell it, is precisely what you need to tell it. Forgive the masculine metaphor, but much depends on this mental tumescence. What I discovered is that it, in turn, depends in no small part on 1,3,7-trimethylxanthine, the tiny organic molecule known to most of us as caffeine.

The first day of my withdrawal, which began on April 10, was by far the most trying, so much so that the prospect of writing, or even just reading, was immediately rendered futile. I had postponed the dark day as long as I could, concocting the kinds of excuses every addict does. "Stressful week coming up," I would inform myself. "Probably not the best time to go cold turkey." Of course, there was never a "good time" to do it—always some reason I needed to be sharp and couldn't afford the "flu-like symptoms" the researchers said might be in store. "I wanna do right," as the country singer Gillian Welch crooned, "but not right now." That was me, day after day. Procrastination at the beginning of any writing project is not unusual for me, but this went on for weeks. Eventually, however, I found myself cornered by the fact that there was no more reporting to be done and that all that stood between me and sitting down to write was quitting coffee—the very act that would render it impossible to write.

I set a date and determined to stick to it.

April 10, a Wednesday morning, arrived. According to the researchers I'd interviewed, the process of withdrawal had actually begun overnight, while I was sleeping, during the "trough" in the graph of caffeine's diurnal effects. The day's first cup of tea or coffee acquires most of its power—*its joy!*—not so much from its euphoric and stimulating properties than from the fact that it is suppressing the emerging symptoms of withdrawal. This is part of the insidiousness of caffeine. Its mode of action, or "pharmacodynamics," mesh so perfectly with the rhythms of the human body, so that the morning cup of coffee arrives just in time to head off the looming mental distress set in motion by yesterday's cup of coffee. Daily, caffeine proposes itself as the optimal solution to the problem caffeine creates. How brilliant!

My morning ritual with Judith—after breakfast and exercise at home—involves a half-mile "walk to coffee," as the real estate brokers now like to say. For some reason we never make coffee at home. Instead, we buy a cup at the Cheese Board, a local bakery and cheese shop, and sip it from a cardboard container swaddled in a warm cardboard sleeve. (Wasteful, I know.) Hoping to fool myself, I made sure to keep everything about the morning ritual unchanged—the walk down the hill and the hot drink in a swaddled cardboard cup—except that when I reached the register I forced myself to ask for a mint tea instead of the usual large half-caf. (Yes, I was a comparative piker in my caffeine consumption.) After years of "the usual," this raised the barista's eyebrow. "I'm on the wagon," I explained, apologetically.

On this morning, that lovely dispersal of the mental fog that the first hit of caffeine ushers into consciousness never arrived. The fog settled over me and would not budge. It's not that I felt terrible—I never got a serious headache—but all day long I felt a certain muzziness, as if a veil had descended in the space between me and reality, a kind of filter that absorbed certain wavelengths of light and sound. I wrote in my notebook, "Consciousness feels less transparent than usual, as if the air is slightly thicker and seems to be slowing everything down, including perception." I was able to do some work, but distractedly. "I feel like an unsharpened pencil," I wrote. "Things on the periphery intrude, and won't be ignored. I can't focus for more than a minute. Is this what it's like to have A.D.D.?"

By noon I was mourning the passing of caffeine from my life for an undetermined period of time. I *so* missed what Judith calls her "cup of optimism"; the same cup that Alexander von Humboldt, the great German naturalist, called "concentrated sunshine." (Humboldt had a parrot named Jacob that could say only one thing: "More coffee, more sugar.") Though at this point I would have settled for much less than optimism. "What I miss," I wrote, "is nothing resembling a state of intoxication or euphoria, just the simple gift of my normal everyday consciousness. Is this my new baseline? God, I hope not."

Over the course of the next few days I definitely began to feel better—the veil lifted—yet I was still not quite myself, and neither, quite, was the world. By the end of the week, I had gotten to the point where I didn't think I could fairly blame caffeine withdrawal for my mental state (and disappointing output), and yet in this new normal the world seemed duller to me. I seemed duller, too. Mornings were the worst. I came to see how integral caffeine is to the daily work of knitting ourselves back together after the fraying of consciousness during sleep. That reconsolidation of self—the daily sharpening of the mental pencil—took much longer than usual and never quite felt complete. I began to think of caffeine as an essential ingredient for the construction of an ego. Mine was now deficient in that nutrient, which perhaps explains why the whole idea of writing this piece—indeed, of ever writing anything ever again—had come to seem insurmountable.

I've been talking here about a chemical—caffeine—but of course we're really talking about a plant, or in this case two plants: *Coffea* and *Camellia sinensis*, aka tea, which, over the course of their evolution, figured out how to produce a chemical that happens to addict most of the human species.* This is an astounding accomplishment, and while that was not the plants' intent in concocting this molecule—there is no intent in evolution, just lots of blind chance that occasionally yields an adaptation so good that it is extravagantly rewarded—once that molecule found its way into the human brain, the destinies of those plant species and this animal species changed in momentous ways.

The adaptation proved so ingenious that it allowed the plants to wildly expand their numbers and habitats. In the case of *Coffea*, whose range had previously been limited to a few corners of East Africa and southern Arabia, its appeal to our species allowed it to circumnavigate the planet, colonizing a broad band of territory, mainly in the tropical highlands, that reaches from Africa to East Asia, Hawaii, Central and South America, and now covers more than 27 million acres. The path of *Camellia sinensis* has taken the plant from its origins in Southwest China (near present-day Myanmar and Tibet) as far west as India and east to Japan, colonizing

---

*A few other plants also produce caffeine, though in smaller amounts, including kola, cacao, yerba mate, guarana, and the yaupon holly, which American southerners have used as a caffeine source when tea and coffee were unavailable.

more than 10 million acres. These are two of the world's most successful plants, right up there with the edible grasses—rice, wheat, and corn. Yet compared to those species, which won our support by so admirably supplying our need for calories, tea and coffee's ticket to world domination involved something much more subtle and superfluous: their ability to change our consciousness in desirable and useful ways. Also unlike the edible grasses, the fat seeds of which we consume with virtually every meal, all we want from the tea and coffee plants are the molecules of caffeine and some characteristic flavors we extract from their leaves and seeds, respectively. So all we do with them is trivially lighten the weight of their vast biomass before simply dumping it all in landfills. Tons of these most valuable of all agricultural commodities are shipped from the tropics to the higher latitudes, there to be briefly soaked in hot water and then thoughtlessly discarded. Isn't there something ecologically absurd about moving all these leaves and seeds around the world merely to inflect water?

Coffee and tea had their own reasons for producing the caffeine molecule, and as is often the case for the so-called secondary metabolites produced by plants, this is for defense against predators. At high doses, caffeine is lethal to insects. Its bitter flavor may also discourage them from chewing on the plants. Caffeine also appears to have herbicidal properties and may inhibit the germination of competing plants that attempt to grow in the zone where seedlings have taken root or, later, dropped their leaves.

Many of the psychoactive molecules plants produce are toxic, but as Paracelsus famously said, the dose makes the poison. What kills at one dose may do something more subtle and interesting at another. The interesting question is why so many of the defense chemicals produced by plants are psychoactive in animals at less-than-lethal doses. One theory holds that the plant doesn't necessarily want to kill its predator, only disarm it. As the long history of the plant defense chemical versus insect arms race demonstrates, killing your predator outright isn't necessarily the best move, since the toxin selects for resistance, rendering it harmless. Whereas if you succeed in merely discombobulating your enemy—distracting him from his dinner, say, or ruining his appetite, as many psychoactive compounds will do—you might be better off, since you will save yourself while preserving the power of your defense toxin.

Caffeine does, in fact, shrink the appetite and discombobulate insect brains. In a famous experiment conducted by NASA in the 1990s, researchers fed a variety of psychoactive substances to spiders to see how they would affect their web-making skills. The caffeinated spider spun a strangely cubist and utterly ineffective web, with oblique angles, openings big enough to let small birds through, and completely lacking in symmetry or a center. (The web was far more fanciful than the ones spun by spiders given cannabis or LSD.) Intoxicated insects are also, like intoxicated humans, more likely to do reckless things, thereby attracting the attention of birds and other predators that will happily do the plant's bidding by snatching and destroying the helplessly dancing or stumbling bug.

Most of the various plant chemicals, or alkaloids, that people have used to alter the textures of consciousness are chemicals originally selected for defense. Yet even in the insect world, the dose makes the poison, and if the dose is low enough, a chemical made for defense can serve a very different purpose: to attract, and secure the enduring loyalty of, pollinators. This appears to be what's going on between bees and certain caffeine-producing plants, in a symbiotic relationship that may have something important to tell us about our own relationship to caffeine.

The story begins in the 1990s, when German researchers made the surprising discovery that several classes of plants—including not only coffee and tea but also the *Citrus* family and a handful of other genera—produce caffeine in their nectar, a substance that evolved to attract rather than repel insects. Was this an accident—a leaking of caffeine from other parts of the plant?—or could it be a slightly diabolical adaptation?

When Geraldine Wright stumbled on the German paper, she was a young lecturer, a botanist turned entomologist, at Newcastle University in England. "We had no idea why caffeine was in the nectar," Wright told me. So in 2013 Wright, who now teaches in the Zoology Department at the University of Oxford, conducted a simple, inexpensive experiment to find out. She trapped a bunch of honeybees and immobilized them in little bee straitjackets, arranging them in a grid of bee-sized roofless apartments with only their heads poking out on top. Using a medicine dropper, Wright fed her bees various mixtures of sugar water with and without different concentrations of caffeine. Each time she offered a bee a drop of pseudonectar, she gave it a little puff of a scent. The idea was to see how quickly the bees learned to associate that scent with a desirable food source.

"Really simple, low-tech, no funding," she said, describing the rudimentary setup. Okay, but how do you determine a bee's food preferences? "That's simple, too," Wright said. "They extend their mouth parts and proboscis if they want something."

Wright discovered that her bees were more likely to remember the odor associated with the caffeinated nectar over the odor associated with sucrose only. (Her results appeared in an article published in *Science* in 2013 called "Caffeine in Floral Nectar Enhances a Pollinator's Memory of Reward.") Even at concentrations too small for the bees to taste, the presence of caffeine helped them to quickly learn and recall a particular scent and to favor it.

You can see why this would be valuable to a flower: it would cause the pollinator to remember that flower and return to it more avidly. Or, as the entomologist put it in the paper, caffeinated nectar increases "pollinator fidelity," otherwise known as floral constancy. Drug your pollinator with a low dose of caffeine and she will remember you and come back for more, choosing you over other plants that don't offer the same buzz.

Actually, we don't know whether the bees feel *anything* when they ingest caffeine, only that the chemical helps them to remember—which, as we will see, caffeine appears to do for us, too. Subsequent experiments with bigger budgets and more elaborate setups, involving fake flowers in more naturalistic settings, have replicated Wright's discovery: bees will remember and return more reliably to flowers that offer them caffeinated nectar. What's more, the power of this effect is so great that bees will continue to return to those flowers even when there is no nectar left. An experiment conducted by Margaret J. Couvillon and published in *Current Biology* in 2015 ("Caffeinated Forage Tricks Honeybees into Increasing Foraging and Recruitment Behaviors") raised the cui bono question: who benefits more from this coevolutionary arrangement between pollinators and caffeine-producing plants? The answer would appear to be the plant.

Couvillon demonstrated that the memory and enthusiasm of the bees for caffeinated flowers was such that it increased "foraging frequency, waggle dancing probability and frequency, and persistency and specificity to the forage location, resulting in a quadrupling of colony-level recruitment"—that is, she estimated that four times as many bees would pay visits to the caffeinated flowers than to flowers offering nectar only. Yet the bees' exuberance exceeds any conceivable benefit to them, making it irrational: "caffeine causes bees to overestimate forage quality, tempting the colony into sub-optimal foraging strategies" likely to "reduce honey storage," since they kept returning to the caffeinated flowers long after they'd been depleted of nectar. She concluded that this makes "the relationship between pollinator and plant less mutualistic and more exploitative." The plant's offer of caffeine to the bees is "akin to drugging, where the pollinator's perception of the forage quality is altered, which in turn changes its individual behaviors." It's an eerily familiar story: a credulous animal duped by a plant's clever neurochemistry into acting against its interests.

An uncomfortable series of questions arises: Could we humans be in the same boat as those hapless bees? Have we, too, been duped by caffeinated plants not only to do their bidding but to act against our own interests in the process? Who's getting the best of our relationship with the caffeine-producing plants?

There are a few different ways to attack this question, but a good one is to attempt to answer two further questions: Has the discovery of caffeine by humans been a boon or a bane to our civilization? And what about to our species, which might not be quite the same thing?

In the case of caffeine, we can look to recorded history for answers, since humanity's acquaintance with caffeine is surprisingly recent. Hard as it is to imagine, Western civilization was innocent of coffee or tea until the 1600s; as it happens, coffee, tea, and chocolate (which also contains caffeine) arrived in England during the same decade—the 1650s—so we can gain some idea of the world before caffeine and after. Coffee was known in East Africa for a few centuries before that—it's believed to have been discovered in Ethiopia around AD 850—but it does not have the antiquity of other psychoactive substances, such as alcohol or cannabis or even some of the psychedelics, like psilocybin or ayahuasca or peyote, which have played a role in human culture for millennia. Tea is also older than coffee, having been discovered in China, and used as a medicine, since at least 1000 BC, though tea wasn't popularized as a recreational beverage until the Tang dynasty, between AD 618 and 907.

It is hardly an exaggeration to say that the arrival of caffeine in Europe changed . . . everything. That sounds hyperbolic, I know, and we often hear something similar about other developments in "material culture"—how the discovery of X or Y (a New World commodity, say, or some invention or discovery) "made the modern world." This usually means that the advent of X or Y had a transformative effect on economics or everyday life or the standards of living. But like the caffeine molecule itself, which rapidly reaches virtually every cell of the body that ingests it, the changes wrought by coffee and tea occurred at a more fundamental level—at the level of the human mind. Coffee and tea ushered in a shift in the mental weather, sharpening minds that had been fogged by alcohol, freeing people from the natural rhythms of the body and the sun, thus making possible whole new kinds of work and, arguably, new kinds of thought, too. Having brought what amounted to a new form of consciousness to Europe, caffeine went on to influence everything from global trade to imperialism, the slave trade, the workplace, the sciences, politics, social relations, arguably even the rhythms of English prose.

The story goes that human engagement with the coffee plant begins with an observant goat herder in present-day Ethiopia, one of a handful of places in Africa where the shrubby tree grows wild. According to the story, a ninth-century herder by the name of Kaldi noticed how his goats would behave erratically and remain awake all night after eating the red berries of the *Coffea arabica* plant. Kaldi shared his observation with the abbot of a local monastery, who concocted a drink with the berries and discovered the stimulating properties of coffee.

Perhaps. But we do know that by the fifteenth century, coffee was being cultivated in East Africa and traded across the Arabian Peninsula. Initially the new drink was regarded as an aid to concentration and used by Sufis in Yemen to keep them from dozing off during their religious observances. (Tea, too, started out as a kind of spiritual NoDoz for Buddhist monks striving to stay awake through long stretches of meditation.) Within a century, coffeehouses had sprung up in cities across the Arab world. In 1570 there were more than six hundred of them in Constantinople alone, and they spread north and west with the Ottoman Empire. These new public spaces were hotbeds of news and gossip, as well as places to gather for performances and games. Coffeehouses were comparatively liberal institutions where the conversation often turned to politics, and at various times governmental and clerical powers-that-be attempted to close them down, but never for long or with much success. (A vat of coffee was put on trial in Mecca in 1511 for its dangerously intoxicating effects; however, its conviction, and subsequent banishment, was quickly overturned by the sultan of Cairo.) As coffee's defenders rightly pointed out, the beverage is nowhere mentioned in the Koran. Coffee thus offered the Islamic world a suitable alternative to alcohol, which is specifically proscribed in the Koran, and it came to be known as *kahve*, which, loosely translated, means "wine of Araby." This notion that coffee somehow exists in opposition to alcohol would persist in both the East and the West, and comes down to us today in the common, but erroneous, belief that black coffee is an antidote for drunkenness.

The Islamic world at this time was in many respects more advanced than Europe, in science and technology, and in learning. Whether this mental flourishing had anything to do with the prevalence of coffee (and prohibition of alcohol) is difficult to prove, but as the German historian Wolfgang Schivelbusch has argued, the beverage "seemed to be tailor-made for a culture that forbade alcohol consumption and gave birth to modern mathematics." In China the popularity of tea during the Tang dynasty also coincided with a golden age. And the far-reaching impact of caffeine's arrival in Europe gives the idea of a causal link some plausibility.

Europeans had long been fascinated by the exotic practices of "the Orient," and the drinking of this inky hot beverage soon sparked their curiosity. A Venetian traveler to Constantinople in 1585 noted that the locals "are in the habit of drinking in public in shops and in the streets, a black liquid, boiling as they can stand it, which is extracted from a seed they call Cave . . . and is said to have the property of keeping a man awake." The notion of drinking any beverage piping hot was itself exotic, and, in fact, this proved to be one of the most important gifts to humanity of both coffee and tea: the fact that you needed to boil water to make them meant that they were the safest things a person could drink. (Before that it had been alcohol, which was more sanitary than water, but not as safe as tea or coffee. The tannins in all these beverages also have antimicrobial properties.) The contribution of coffee and tea to public health may help explain why societies that embraced the new hot drinks tended to thrive, as microbial diseases declined.

In 1629 the first coffeehouses in Europe, styled on the Arab model, popped up in Venice, and the first such establishment in England was opened in Oxford in 1650 by a Jewish immigrant known as Jacob the Jew. They arrived in London shortly thereafter, and proliferated virally: within a few decades there were thousands of coffeehouses in London; at their peak, one for every two hundred Londoners.

As in the Islamic world, in Europe coffee was mainly consumed in public coffeehouses—vibrant meeting places where the news of the day (political, financial, and cultural) was as much the draw as the coffee. Coffeehouses became uniquely democratic public spaces; in England they were the only such spaces where men of different classes could mix. Anyone could sit anywhere. But *only* men, at least in England, a fact that led one wag to warn that the popularity of coffee "put the whole race in danger of extinction." (Women were welcome in French coffeehouses.) Compared to taverns, coffeehouses were also notably civil places where, if you started an argument, you were expected to buy a round for everyone.

To call the English coffeehouse a new kind of public space

doesn't quite do it justice; it represented a new kind of communications medium, one that just happened to be made of brick and mortar rather than electricity and wires. You paid a penny for the coffee, but the information—in the form of newspapers, books, magazines, and conversation—was free. (Coffeehouses were often referred to as "penny universities.") After visiting London coffeehouses, a French writer named Maximilien Misson wrote, "You have all Manner of News there; You have a good fire, which you may sit by as long as you please: You have a Dish of Coffee; you meet your Friends for the Transaction of Business, and all for a Penny, if you don't care to spend more."

London's coffeehouses were distinguished one from another by the professional or intellectual interests of their patrons, which eventually gave them specific institutional identities. So, for example, merchants and men with interests in shipping gathered at Lloyd's Coffee House. Here you could learn what ships were arriving and departing, and buy an insurance policy on your cargo. Lloyd's Coffee House eventually became the insurance brokerage Lloyd's of London. Similarly, the London Stock Exchange had its roots in the trades conducted at Jonathan's Coffee-House. Learned types and scientists—known then as natural philosophers—gathered at the Grecian, which became closely associated with the Royal Society; Isaac Newton and Edmund Halley debated physics and mathematics here, and supposedly once dissected a dolphin on the premises. Tom Standage, author of *A History of the World in 6 Glasses* (three of which happen to contain caffeine: coffee, tea, and cola), writes that coffeehouses "provided an entirely new environment for social, intellectual, commercial, and political exchange," making those in

London what he calls "the crucibles of the scientific and financial revolutions that shaped the modern world."

Meanwhile, the literary set gathered at Will's and at Button's, in Covent Garden, where you might bump into John Dryden or Alexander Pope. Pope's "The Rape of the Lock" is steeped in the culture, and particularly the gossip, of the coffeehouse, and, in Canto III, pays homage to the power of the brew, "which makes the politician wise." It also supplied an important plot point: It was coffee that "Sent up in vapours to the Baron's brain / New Stratagems, the radiant Lock to gain." Some critics maintain that the culture of the coffeehouse altered English prose in enduring ways. Habitués like Henry Fielding, Jonathan Swift, Daniel Defoe, and Laurence Sterne brought the rhythms of spoken English into their prose, marking a radical turn from the formality of previous English prose stylists.

Specialized though they were by field of interest, London's coffeehouses were also linked by patrons who spent the day moving from one to another, carrying news but also rumors and gossip, which spread more quickly through London's network of coffeehouses than by any other medium.

One of England's earliest magazines, *The Tatler*, began its life in the Grecian in 1709 and was itself an attempt to translate the sheer variety of London's coffeehouse culture to the page. The magazine was divided into sections, each covering a different subject and named for the coffeehouse associated with that particular interest. As Richard Steele, the *Tatler*'s editor, explained in an early issue, "All accounts of Gallantry, Pleasure, and Entertainment shall be under the Article of White's Chocolate-house; Poetry, under that of

Will's Coffee-house; Learning, under the title of Graecian; Foreign and Domestick News you will have from St. James' Coffee-house."

Not everyone in seventeenth-century England approved of coffee or of the coffeehouse. Medical men debated the beverage's healthfulness in fevered tracts, and women strenuously objected to the amount of time men were spending in coffeehouses. In a pamphlet titled "The Women's Petition Against Coffee" published in 1674, the authors suggested that the "Enfeebling Liquor" robbed men of their sexual energies, making them "as unfruitful as those Desarts whence that unhappy Berry is said to be brought."

The unsubtle subtitle of the pamphlet—"Humble Petition and Address of Several Thousands of Buxome Good Women, Languishing in Extremity of Want"—did not mince words: men were spending so much time in coffeehouses, and drinking so much coffee, that they arrived home with "nothing stiffe but their joints." The men replied with their own pamphlet, claiming that the "Harmless and healing liquor . . . makes the erection more Vigorous, the Ejaculation more full, [and] adds a spiritualescency to the Sperme." Any problem in this department the pamphleteers wrote off to the "Husband's natural infirmity" or possibly "your own perpetual Pumping him, not drinking coffee."

The seventeenth-century war of the sexes over coffee led to the association of tea with femininity and domesticity that endures to this day in the West. A Londoner could get a cup of tea in the coffeehouse, but tea didn't have its own dedicated public venue until 1717, when Thomas Twining opened a tea house next door to Tom's, his coffeehouse in the Strand. Here women were welcome to sample the various offerings and buy tea leaves to brew at home. Thanks in part to Twining's innovation, what was soon to become the more popular caffeinated beverage in Great Britain came under the control of upper- and middle-class women, who proceeded to develop a rich culture of tea parties, high teas and low, and a whole regime of tea accessories, including china and porcelain, the teaspoon and the tea cozy, and finger foods expressly designed to accompany tea. (The temperance movement, led by women and promoting tea as an alternative to gin, would later solidify tea's feminine image in the West.)

Women's were not the only voices raised against coffee drinking. The conversation in London's coffeehouses frequently turned to politics, in vigorous exercises of free speech that drew the ire of the government, especially after the monarchy was restored in 1660. Charles II, worried that plots were being hatched in coffeehouses, decided that the places were dangerous fomenters of rebellion that the Crown needed to suppress. In 1675 the king moved to close down the coffeehouses, on the grounds that the "false, malicious and scandalous Reports" emanating therefrom were a "Disturbance of the Quiet and Peace of the Realm." Like so many other compounds that change the qualities of consciousness in individuals, caffeine was regarded as a threat to institutional power, which moved to suppress it, in a foreshadowing of the wars against drugs to come.

But the king's war against coffee lasted only eleven days. Charles discovered that it was too late to turn back the tide of caffeine: by then the coffeehouse was such a fixture of English culture and daily life—and so many eminent Londoners had become dependent upon caffeine—that everyone simply ignored the king's order and blithely went on drinking coffee. Afraid to test his authority and find it lacking, the king quietly backed down, issuing a second proclamation rolling back the first "out of princely consideration and royal compassion."

In France, too, coffeehouses became synonymous with sedition, and would play a decisive role in the events of 1789. Jules Michelet wrote that those "who assembled day after day in the Café de Procope saw, with a penetrating glance, in the depths of their black drink, the illumination of the year of the revolution." Perhaps for this reason, Paris's coffeehouses were rife with intrigue. The mob that ultimately stormed the Bastille assembled in the Café de Foy, roused to action by the eloquence of political journalist Camille Desmoulins and intoxicated not by alcohol but by caffeine.

It's hard to imagine that the sort of political, cultural, and intellectual ferment that bubbled up in the coffeehouses of both France and England would ever have developed in a tavern. If alcohol fuels our Dionysian tendencies, caffeine nurtures the Apollonian. Early on, people recognized the link between the rising tide of rationalism and the fashionable new beverage. "Henceforth is the tavern dethroned," Michelet wrote, surely overstating the case. Wine and beer did not go away, yet the European mind had been pried loose from alcohol's grip, freeing it for the new kinds of thinking that caffeine helped to foster. You can argue what came first, but the kind of magical thinking that alcohol sponsored in the medieval mind began in the seventeenth century to yield to a new spirit of rationalism and, a bit later, Enlightenment thinking. Continues Michelet: "Coffee, the sober drink, the mighty nourishment of the brain, which unlike other spirits, heightens purity and lucidity; coffee, which clears the clouds of the imagination and their gloomy weight; which illumines the reality of things suddenly with the flash of truth." To see, lucidly, "the reality of things": this was, in a nutshell, the rationalist project. Coffee became, along with the microscope, the telescope, and the pen, one of its indispensable tools. But unlike the others, this was a tool that was taken up *in* the brain and mind. Wolfgang Schivelbusch writes in his wonderful history of stimulants and intoxicants, *Tastes of Paradise*, "With coffee, the principle of rationality entered human physiology, transforming it to conform with its own requirements."

The enthusiasm for coffee among intellectuals in both England and France reflected, perhaps, its novelty as much as its power: new drugs always seem miraculous, and for that reason are often credited with astounding properties and consumed to excess. Voltaire was a fervent advocate for coffee, and supposedly drank as many as seventy-two cups a day. Coffee, and coffeehouses, fueled heroic labors in Enlightenment writers. Denis Diderot compiled his magnum opus while imbibing caffeine at the Café de Procope. It's safe to say the *Encyclopédie* would never have gotten finished in a tavern.

Honoré de Balzac was convinced his vast literary output, as well as the operations of his imagination, depended on heroic doses of coffee, consumed through the night as he chronicled the human comedy in his innumerable novels. Eventually, he developed such a tolerance for caffeine that he dispensed altogether with the diluting effects of water, developing his own unique method of administering the drug dry:

> I have discovered a horrible, rather brutal method that I recommend only to men of excessive vigor. It is a question of using finely pulverized, dense coffee, cold and anhydrous, consumed on an empty stomach. This coffee falls into your stomach, a sack whose velvety interior is

> lined with tapestries of suckers and papillae. The coffee finds nothing else in the sack, and so it attacks these delicate and voluptuous linings . . . sparks shoot all the way up to the brain.

The effect for Balzac was to transform the brain into a pitched mental battleground where the epic forces of his imagination could contend:

> From that moment on, everything becomes agitated. Ideas quick-march into motion like battalions of a grand army to its legendary fighting ground, and the battle rages. Memories charge in, bright flags on high; the cavalry of metaphor deploys with a magnificent gallop, the artillery of logic rushes up with clattering wagons and cartridges; on imagination's orders, sharpshooters sight and fire; forms and shapes and characters rear up; the paper is spread with ink . . .

Perhaps not surprisingly it was Balzac who wrote one of the all-time best descriptions of how it feels to be overcaffeinated, a state that he said

> produces a kind of animation that looks like anger: one's voice rises, one's gestures suggest unhealthy impatience; one wants everything to proceed with the speed of ideas; one becomes brusque, ill-tempered, about nothing. One assumes that everyone else is equally lucid. A man of spirit must therefore avoid going out in public.

It is one thing to live in a shared culture of caffeine, in which everyone's mind is running at more or less the same accelerated pace. But it's quite another to find yourself so sped up mentally that other people appear to you like motionless figures on a train platform, as you blur by them in caffeinated clouds of impatience.

Balzac's account of caffeine intoxication hit home as I approached the third month of my abstention. I felt very much like that stationary figure on the platform, catching envious glimpses of coffee drinkers through the train window as they streaked by.

After a few weeks, the mental impairments of withdrawal had subsided, and I could once again think in a straight line, hold an abstraction in my head for more than two minutes, and shut peripheral thoughts out of my field of attention. My confidence in telling this story gradually returned, and after a month I could write again; you can judge how well that's going, but at least it's going. Yet I continue to feel as though I'm mentally just slightly behind the curve, especially when in the company of drinkers of coffee and tea, which, of course, is all the time and everywhere. In college I dated a woman who had grown up without a television in her home; she missed so many references, jokes, and allusions that she sometimes seemed vaguely foreign to us, and we to her. There was this subtle but unmistakable mental hitch. These days, it feels a little like that.

Here's what I'm missing: I miss the way caffeine and its rituals used to order my day, especially in the morning. Herbal teas—which are barely, if at all, psychoactive—lack coffee's and tea's power to organize the day into a rhythm of energetic peaks and valleys, as the mental tide of caffeine ebbs and flows. The morning surge is a blessing, obviously, but there is also something comforting in the ebb tide of afternoon, which a cup of tea can gently reverse.

I miss the enveloping aroma and the sounds of coffee, whether it's the mechanical scream of beans being ground or the more contented burbling of the coffee as it percolates. Actually, those sensory gifts are still available to me—every time I walk past a café—but it turns out the smells and sounds by themselves are merely a taunt if not followed by consummation. Lately I've taken to brewing coffee for Judith at home, grinding the smoky-woodsy aromas out of the beans and inhaling the steam off her cup before I hand it to her, hoping to imbibe some hint of mental stimulant before I head off to my desk to sip my chamomile tea. What work of genius has ever been composed on chamomile? What mental breakthrough has ever been credited to mint tea? It's a miracle I've gotten this far with our story.

I miss being able to take part in coffee culture, idling in cafés and taking in the scene. Even as the mind accelerates, the body slows and is perfectly content to while away the time. Curiously, that culture no longer revolves around conversation, which has all but dried up in the modern coffeehouse; it's been replaced by the mental industry of coffee drinkers tapping away on their laptops, with a sense of urgency that I can't even pretend to possess. So many important projects! Sure, I can sit there among them with my tisane, but it's not the same. I no longer swim in the same caffeine sea as everyone else. Beached, I can still see the water—but it's way over there.

There are some compensatory benefits. I'm sleeping like a teenager again, and wake feeling actually refreshed. (There's an explanation for this I will get to.) I've discovered an odd and unexpected social benefit as well. When I turn down offers of coffee and explain my experiment in abstention, I find that people are keenly interested and, oddly, sort of impressed. It's as though I've notched some kind of achievement. "I could never do that," a friend will say, or, "I should really try that; I know it would help me sleep. But I can't imagine getting through the morning." Naturally, these reactions make me feel as though I've actually accomplished something worthy of admiration. I suspect I'm benefiting from the echoes of Puritanism still reverberating in our culture, which even now awards points for self-discipline and overcoming desire. Addiction, even to a relatively harmless and easily procured drug like caffeine, is seen as evidence of weakness of character. "I realized my life was being controlled by caffeine," a sleep researcher (and caffeine abstainer) I interviewed told me. "Traveling, I'd find myself in an unfamiliar city and could not turn in for bed until I had scoped out where I was going to get my fix in the morning. I like to feel in control and realized I wasn't. Caffeine was controlling me."

Roland Griffiths, the drug researcher, told me that he had been inspired to study caffeine after embarrassing himself with his own "revolting behavior." In a hurry and in need of a caffeine fix, he had thrown some frozen coffee grounds into a cup, added hot tap water, swished it around, and downed it. "I recognize drug-seeking behavior when I see it!" Yet he agreed that there's nothing inherently "wrong" with an addiction if you have a secure supply, no known health risk, and you're not offended by the idea. But many of us can't help moralizing addiction.

I will confess to indulging in the occasional pang of righteousness. Ordinarily a walk through the airport during my months of abstention filled me with yearning and envy, as I rolled past one aromatic caffeine opportunity after another. But matters look very different to a reformed addict first thing in the morning. One such morning, having hauled myself out of bed and to the airport for a 6:00 a.m. flight, fueled on nothing but peppermint tea, I registered only pity as I beheld the lines snaking in front of the Starbucks and

Peet's, lines so long it would take easily a half hour for these poor wretches to get served. I could see they were enduring the first symptoms of caffeine withdrawal, and their desperation to head them off and return to baseline consciousness carried a whiff of pathos. They looked like better-dressed versions of the addicts I had seen in Amsterdam, lining up in front of a mobile dispensary for their morning fix. I thought to myself, *These people are pathetic!* This is not a thought I am proud of; in fact, I look forward to rejoining the ranks of the caffeine-dependent as soon as I can. In the meantime, however, I try to savor the moral elevation and self-regard one feels living free of this addiction. For now, that's pretty much all I've got.

At some point I began to wonder if perhaps it was all in my head, this sense that I had lost a mental step since getting off coffee and tea. The debt to coffee so freely acknowledged by the mental giants of the age of reason and Enlightenment fed my suspicion that I might still be suffering from a subtle, or perhaps not so subtle, mental deficiency. Since I had not given up wine during the period of my caffeine abstention, was it possible that I had personally reversed the forward march of intellectual progress in the West, casting myself back into the medieval mists of slow and magical thinking? Yet even without the mental clarity bestowed by caffeine, I knew better than to put much weight on the anecdotal or the sample of 1. So I decided to check in with Science, to learn what, if any, cognitive enhancement can actually be attributed to caffeine. What was I really missing?

I found numerous studies conducted over the years reporting that caffeine improves performance on a range of cognitive measures—of memory, focus, alertness, vigilance, attention, and learning. An experiment done in the 1930s found that chess players on caffeine performed significantly better than players who abstained. In another study, caffeine users completed a variety of mental tasks more quickly, though they made more errors; as one paper put it in its title, people on caffeine are "faster, but not smarter." In a 2014 experiment, subjects given caffeine immediately after learning new material remembered it better than subjects who received a placebo. Tests of psychomotor abilities also suggest that caffeine gives us an edge: In simulated driving exercises, caffeine improves performance, especially when the subject is tired. It also enhances physical performance on such metrics as time trials, muscle strength, and endurance.

True, there is reason to take these findings with a pinch of salt, if only because this kind of research is difficult to do well. The problem is finding a good control group in a society in which virtually everyone is addicted to caffeine. If you compare the performance of two groups, one to whom you've given a caffeine tablet and the other a placebo, the chances are strong that the placebo group is in the throes of caffeine withdrawal, and so at a distinct disadvantage performing any sort of cognitive or motor task. It could be that the caffeine is merely restoring volunteers to normal baseline mental function rather than enhancing it.

Researchers can overcome this problem by making sure their volunteers have been free of caffeine for a week or two, and many of them do. The consensus seems to be that caffeine does improve mental (and physical) performance to some degree. The science suggests that in all likelihood I *have* lost a mental step since embarking on this experiment, relative to my previous coffee- and tea-drinking self. I hereby apologize for any lapses that might have occurred as a result.

Whether caffeine also enhances creativity is a different question, however, and there's some reason to doubt that it does, Balzac's fervent belief to the contrary. Caffeine improves our focus and ability to concentrate, which surely enhances linear and abstract thinking, but creativity works very differently. It may depend on the *loss* of a certain kind of focus, and the freedom to let the mind off the leash of linear thought.

Cognitive psychologists sometimes talk in terms of two distinct types of consciousness: spotlight consciousness, which illuminates a single focal point of attention, making it very good for reasoning, and lantern consciousness, in which attention is less focused yet illuminates a broader field of attention. Young children tend to exhibit lantern consciousness; so do many people on psychedelics. This more diffuse form of attention lends itself to mind wandering, free association, and the making of novel connections—all of which can nourish creativity. By comparison, caffeine's big contribution to human progress has been to intensify spotlight consciousness—the focused, linear, abstract, and efficient cognitive processing more closely associated with mental work than with play. This, more than anything else, is what made caffeine the perfect drug not only for the age of reason and the Enlightenment but for the rise of capitalism, too.

Speaking of focus . . . sorry, but I didn't mean to drop the thread of caffeine history we had been following awhile back. Let me try to pick it up.

The soaring popularity of the coffeehouse in seventeenth-century Europe posed a problem for business interests there since, at the time, Arab traders had an absolute monopoly on coffee beans; they profited from every cup of coffee consumed in London, Paris, or Amsterdam. It was a monopoly the Arabs zealously guarded: to prevent anyone from growing coffee anywhere but in the lands they controlled, Arab traders roasted coffee beans (which are seeds, after all) before they were exported, to ensure they could not be germinated.

But in 1616, a wily Dutchman managed to break the Arab stranglehold on *Coffea arabica*. He smuggled live coffee plants out of Mocha, the Yemeni port city, and took them to the botanical garden in Amsterdam, where they were grown under glass and additional plants were eventually propagated by cutting. (You can create a new, genetically identical plant by rooting a shoot or branch in soil.) One of those clones ended up in the Dutch-controlled Indonesian island of Java, where the Dutch East India Company successfully propagated it, eventually producing enough coffee plants to establish a plantation there. Hence, the prized coffee known as Mocha Java.

In 1714 two descendants of the Dutchman's larcenous coffee bush were given to King Louis XIV, who had it planted in the Jardin du Roi, in Paris. A few years later, a former French naval officer named Gabriel de Clieu dreamed up a scheme to establish coffee production in the French colony of Martinique, where he lived. In a second momentous coffee theft, he claimed to have recruited a woman at court to purloin a cutting of the king's plant.

After successfully rooting the cutting, de Clieu installed the little plant in a glass box to protect it from the elements and brought it with him on a ship bound for Martinique. The crossing proved difficult, taking so much longer than anticipated that the supply of drinking water on board had to be strictly rationed. Determined to keep his coffee plant alive, de Clieu shared his meager ration of water with it.

De Clieu claimed to have nearly died of thirst at sea, but his sacrifice ensured that the plant made it safely to Martinique, where it thrived. By 1730, France's Caribbean colonies were shipping coffee back to what by then was a Europe hopelessly addicted to caffeine. Many of the coffee plants grown in the New World today are descendants of that original plant smuggled out of Mocha in 1616, offspring of a theft nearly Promethean in its impact. Now the West had taken control of coffee—and coffee took control of the West.

Before the arrival of coffee and tea, alcohol was being consumed in Europe morning, noon, and night; not only in taverns after dark but for breakfast at home and even in the workplace, where it was routinely given to laborers on their breaks. The English mind in particular was befogged most of the day by more or less constant infusions of alcohol. Campaigns for temperance sprang up from time to time, but without a substitute beverage they failed to gain traction.

Enter coffee.

As early as 1660, writer and historian James Howell could note: "'Tis found already, that this Coffee drink hath caused a greater Sobriety among the Nations; for whereas formerly Apprentices & Clerks with others used to take their mornings' draught in Ale, Beer, or Wine, which by the dizziness they cause in the Brain, make many unfit for business, they use now to play the Good-fellows in this *wakeful* and civil drink."

Howell deserves credit for having recognized so early the impact of coffee on the conduct of work, for it would prove far-reaching years later when the English economy would begin its shift from a reliance on physical labor to mental labor. Long before the coffee break there was the beer break, commonly offered to laborers doing physical work outdoors; mental clarity was not a priority, nor was attention to clock time. For laborers working with machines, however, a mind dulled by alcohol was a hazard to both safety and productivity. And for clerks and others who worked with numbers, the alertness, focus, and all-around mental clarity coffee afforded made it the ideal drug—"the beverage of the modern bourgeois age," in the words of Wolfgang Schivelbusch. Coffee showed up in Europe at exactly the right moment: "It spread through the body and achieved chemically and pharmacologically what rationalism and the Protestant ethic sought to fulfill spiritually and ideologically." The rationalist drug par excellence, coffee helped disperse Europe's alcoholic fog, fostering a heightened alertness and attention to detail, and, as employers soon discovered, dramatically improving productivity.

Surely it is more than a coincidence that caffeine and the minute hand on clocks arrived at more or less the same historical moment. For medieval man, and especially for the man doing physical labor outdoors, the angle of the sun mattered more than the hand of the clock. There had been no minute hand because there had been no need to subdivide the hour. But new kinds of work demanded much closer attention to time and its increments, and what psychoactive drug is more time-bound than caffeine? Is more closely tied to the temporal landmarks of the day? (Think of T. S. Eliot's Prufrock, measuring out his life in coffee spoons.) Work now was not only moving indoors but also being reorganized on the principle of the clock, regularized and routinized, and this shift called for a new temporal discipline that coffee and tea could help to enforce.

But the most important contribution that caffeine made to modern work—and, in turn, to the rise of capitalism—was to liberate us from the fixed rhythms of the sun, an astronomical timepiece that also sets the clocks of our bodies. Before caffeine, the whole idea of a late shift, let alone a night shift, was inconceivable—the human body simply would not permit it. But the power of caffeine to keep us awake and alert, to stem the natural tide of exhaustion, freed us from the circadian rhythms of our biology and so, along with the advent of artificial light, opened the frontier of night to the possibilities of work. This "wakefulness wrested from Nature," as one early-nineteenth-century German physician described caffeine's gift to humankind, thus allowed us to adapt our bodies and our minds to the requirements of modern life.

And industry. What coffee did for clerks and intellectuals, tea would soon do for the English working class. Indeed, it was tea from the East Indies—heavily sweetened with sugar from the West Indies—that fueled the Industrial Revolution. We think of England as a tea culture, but coffee, initially the cheaper beverage by far, dominated at first. It wasn't until the British East India Company (which had limited access to coffee-producing regions) began trading regularly with China in the first part of the eighteenth century that tea could displace coffee as the principal medium for delivering caffeine to the English bloodstream.

The story of tea has a completely different complexion in the East and the West, suggesting that the meanings we attribute to these psychoactive plants owe as much to the cultural context in which they are consumed as to their inherent qualities, although those surely figure, too. In the East, tea was less about labor and commerce than it was an instrument of the spiritual life, beginning in Taoism and Confucianism and culminating in Zen Buddhism.

The first tea plantations in China were cultivated thousands of years ago by monks, who found that sipping tea was an important aid to meditation. In one of the origin stories for the discovery of tea, Bodhidharma, a sixth-century Indian prince seeking enlightenment, was in the midst of a seven-year-long meditation (he had already completed a nine-year stint sitting in front of a wall "listening to the ants scream") when, despite his determination to stay awake, he fell asleep. Furious with himself, Bodhidharma cut off his eyelids and threw them on the ground. Tea bushes sprouted where his eyelids landed, a plant with leaves resembling eyelids. From that time forward, the drink would help monks stay awake during the long hours of meditation.

Tea was celebrated in China and, later, Japan not only as a promoter of wakefulness but of health, too—and with good reason. Tea was used as a mouthwash in the East long before science discovered it contains fluoride (the English would negate this advantage by adding copious amounts of sugar to their tea); tea also contains a great many vitamins and minerals—one of the highest concentrations in any plant—and prodigious quantities of polyphenols, compounds rich in antioxidants. (Tea contains more polyphenols than red wine.)

"Always sip tea as if tea were life itself": this injunction, from the eighth-century text *Ch'a-ching*, or *The Classic of Tea*, hints at the exalted role tea played in the spiritual life of China and Japan. The subtleties of this delicate inflection of water, in taste and aroma and appearance, encouraged precisely the kind of concentration and attention to the present moment that Buddhism sought to instill.

The idea that the act of sipping tea could be a spiritual practice culminated in the Zen tea ceremony. Here the scrupulous attention to every physical gesture and material detail gave participants an opportunity to step outside the bustle and messiness of daily life, turning their minds instead to the Zen principles of reverence, purity, harmony, and tranquility. Approached in this spirit of transcendence, the tea ceremony held the power to change consciousness. As the seventeenth-century Japanese tea master Sen Sotan put it, "The taste of tea and the taste of Zen are the same."*

Tea lost most of that taste on its transit from East to West, which transformed it from an instrument of spirituality into a commodity. This shift began as a by-product of the spice trade. There was no demand for tea in Europe when traders scouring the East for spices began adding a few chests of tea to their cargoes. They had no idea this afterthought would soon become a far more important item of trade than spice was and, in time, the most popular beverage on the planet.

Soon after the British East India Company began trading with China, cheap tea flooded England, rapidly displacing coffee as the nation's preferred caffeine delivery system. A beverage that only the well-to-do could afford to drink in 1700 was by 1800 consumed by virtually everyone, from the society matron to the factory worker. To supply this demand required an imperialist enterprise of enormous scale and brutality, especially after the British decided it would be more profitable to turn India, its colony, into a tea producer, than to buy tea from the Chinese. This required first stealing the secrets

*For more on the tea ceremony and, more generally, the place of tea in the spiritual life of China and Japan, see Beatrice Hohenegger, *Liquid Jade: The Story of Tea from East to West* (New York: St. Martin's Press, 2006).

of tea production from the Chinese (a mission accomplished by the renowned Scots botanist and plant explorer Robert Fortune, disguised as a Mandarin), seizing land from peasant farmers in Assam (where tea grew wild), and then forcing the farmers into servitude, picking tea leaves from dawn to dusk.* The introduction of tea to the West was all about exploitation—the extraction of surplus value from labor, not only in its production in India but in its consumption in England as well.

In England, tea allowed the working class to endure long shifts, brutal working conditions, and more or less constant hunger; the caffeine helped quiet the hunger pangs, and the sugar in tea became a crucial source of calories. (From a strictly nutritional standpoint, workers would have been better off sticking with beer.) But in addition to helping capital extract more work from labor, the caffeine in tea helped create a new kind of worker, one better adapted to the rule of the Machine—demanding, dangerous, and incessant. It's difficult to imagine an Industrial Revolution without it.†

I've avoided, at least up to now, attempting to answer the questions of value with which we began, when I wondered whether caffeine represented a boon or a bane to civilization and/or our species.

The widespread use of caffeine is, arguably, one of those developments in human history, like the control of fire or the domestication

---

*This is backbreaking work, which, to this day, is performed mostly by hand. A tea picker is expected to pick up to 30 kilos of leaves per day, requiring 60,000 snips, each of a bud and 2 tea leaves.

†The story of tea unfolds somewhat differently in the American colonies. As Englishmen, the colonialists acquired the tea habit around the same time as their countrymen. But in the eighteenth century, they rebelled at the high taxes the king levied on tea, in one of the first acts in the drama of the Revolution. On December 16, 1773, protesters dumped 342 tea chests—containing 120,000 pounds of tea—into Boston Harbor. After the Boston Tea Party, the patriotic beverage became coffee, which ever since has been more popular than tea in the United States.

of plants and animals, that helped lift us out of the state of nature, providing a new degree of control over biology, in this case our own. But is this an absolutely good or bad thing?

I put the question of whether caffeine was a boon or not to Roland Griffiths during one of our Skype interviews. He had a tall Starbucks cup in front of him, and paused for a long time before answering. "Sure, given the way our culture works, that we have times we need to be awake and asleep and need to report to work at certain times. We're no longer able to just respond to our natural biological rhythms, so to the extent that caffeine helps us sync up our rhythms to the requirements of civilization, caffeine is useful. Whether that's helpful to our species is another question," he finished, trailing off, but clearly implying it was not.

Much depends on where you stand on the trade-offs of modern life and, especially, those of capitalism. Philosopher Michel Foucault's concept of "body discipline" could profitably be used to describe the effects of caffeine, since it helped bend humans to the wheel of the Machine and the requirements of a new economic and mental order. Looked at that way, caffeine is a curse, addicting us to a regime that makes us more tractable and productive workers, speeding us up so that we may better keep pace with the manmade machinery of modern life.

The question of who benefited more from the advent of caffeine, factory or worker, capital or labor, was the subject of a lively debate that came to a head in mid-twentieth-century America. In the 1920s, a time when management and efficiency emerged as a scientific discipline, the impact of coffee on the workplace was studied closely. A consensus emerged that it led to an "increased capacity for work," in the words of one researcher, Charles W. Trigg, and offered "an aid to factory efficiency." But scientists were perplexed as to exactly *how* caffeine could augment people's energy. Energy in biological systems was understood as a function of calories, yet unsweetened coffee or tea contained no calories whatsoever. So where did this new increment of human energy come from? It seemed to violate the laws of thermodynamics, suggesting that caffeine might offer a kind of physiological free lunch. But regardless of whether this could be explained scientifically, employers were quick to recognize and seize on the potential benefit of caffeine—to themselves.

(Actually, one of the first American "employers" to seize on the practical value of caffeine was the Union Army during the Civil War. The army issued each soldier thirty-six pounds of coffee a year at the same time the economic blockade of the South deprived the Confederacy of coffee. According to historian Jon Grinspan, the loss of coffee took a toll on the morale—and perhaps also the performance—of Confederate soldiers, while its easy availability to Union soldiers gave them an edge. One Union general went so far as to weaponize caffeine, ordering his soldiers to fill their canteens with coffee before battle and planning his attacks for the times when his troops were maximally caffeinated. But the amped troops symbolized a larger truth: that the Civil War represented the victory of the caffeinated North, with its sped-up industrialized economy, over the slower, uncaffeinated economy of the Confederacy. Ever since, the American military has made caffeine in all its forms—including tablets and a specially formulated chewing gum—readily available to its soldiers.)

To better understand the origins of the "coffee break," a term that doesn't enter the vernacular until the 1950s, consider the case of two companies in Buffalo, New York—the Larkin Company (a soap manufacturer) and the Barcalo Manufacturing Company (maker of the BarcaLounger) in the first years of the twentieth century. Barcalo offered midmorning and midafternoon breaks to employees; however, they had to bring in and brew their own coffee. (Workers chipped in to buy the coffee, and the company's sole female employee brewed it.) Larkin, by contrast, offered coffee free to its employees, but didn't give them any break time during which to drink it.

It wasn't until the 1950s that the modern concept of the coffee break—free coffee plus paid time in which to enjoy it—was fully established as a legally recognized institution in the American workplace. This happened at a neckwear company in Denver called Los Wigwam Weavers. (The story is told in the 2020 book *Coffeeland*, by historian Augustine Sedgewick.) When Wigwam owner Phil Greinetz lost his best young employees to the war effort, he hired older men to operate the looms. Because of the intricacy of the designs and the number of colors in the neckties, the work was exacting and exhausting, and the older men failed to meet the company's quality standards. Greinetz then tried hiring middle-aged women to operate the looms. The women had the necessary dexterity but lacked the endurance to work a full shift. At a company-wide meeting to discuss the problem, employees proposed that they be given two fifteen-minute breaks—one in the morning, one in the afternoon—and that they be provided with coffee.

Greinetz took their suggestion, establishing a break room and supplying it with coffee and tea. Very soon he "noticed a change in his workers," Sedgewick writes. "Four women who had been among the worst workers were now among the best. Altogether the middle-aged women began to do as much work in six and a half hours as the older men had done in eight hours. Encouraged, Greinetz made the breaks compulsory."

Yet Greinetz felt he shouldn't have to pay the workers for what he regarded as time off, so he docked them for the thirty minutes of break time. Deducting this time from the employees' paychecks caused their wages to fall below the federal minimum, however, prompting a suit against the company from the U.S. Department of Labor. "In court," Sedgewick writes, "Greinetz testified to the extraordinary changes he had observed in his employees" since instituting the coffee breaks, but because the breaks weren't work time, he argued, he wasn't obliged to pay his workers for it.

The company ultimately lost in federal court. The court ruled that though the breaks certainly benefited the workers, they were at least "equally beneficial to the employer in that they promote more efficiency and result in a greater output, and this increased production is one of the primary factors, if not the prime factor, which leads the employer to institute such break periods." The judge also pointed out, rightly, that coffee breaks bore "a close relationship" to work and thereby must be compensated as such. The decision enshrined the paid coffee break in American life. As Sedgewick points out, "the principle that physiologists and bosses had already discovered in practice—that coffee adds something to the working power of the human body independent of the processes and timescales of eating and digestion, something beyond what the science of energy and laws of thermodynamics say is possible—became itself a kind of law."

As for the term "coffee break," it appears to have been popularized in 1952, in an advertising campaign by the Pan-American

Coffee Bureau, the marketing arm of coffee growers in South and Central America. Their slogan: "Give yourself a coffee-break . . . and get what coffee gives to you!"

So how exactly does coffee, and caffeine more generally, give us what it gives us? How could this little molecule possibly supply the human body energy without calories? Could caffeine be the proverbial free lunch? Or do we pay a price for the mental and physical energy—the alertness, focus, and stamina—that caffeine gives us?

To answer these questions, it's necessary to understand something about the pharmacology of caffeine. Caffeine is a tiny molecule that happens to fit snugly into an important receptor in the central nervous system, allowing it to occupy it and therefore block the neuromodulator that would normally bind to that receptor and activate it. That neuromodulator is called adenosine; caffeine, its antagonist, keeps adenosine from doing its job by getting in its way.

Adenosine is a psychoactive compound that has a depressive and hypnotic (that is, sleep-inducing) effect on the brain when it binds to its receptor. It diminishes the rate at which our neurons fire. Over the course of the day, adenosine levels gradually rise in the bloodstream, and as long as no other molecule is blocking its action, it begins to slow mental operations in preparation for sleep. As adenosine builds up in your brain, you begin to feel less alert and a mounting desire to go to bed—what scientists call sleep pressure.

But when caffeine beats adenosine to those receptor sites, the brain no longer receives the signal to begin turning out the mental lights. Even so, the adenosine is still circulating in your brain—in fact, its levels continue to rise—but because the receptors have been hijacked, you don't feel its effects. Instead, you feel wide-awake and alert. Are you really? Yes and no. How you feel is how you feel, it's true, but as Matthew Walker, a UC Berkeley neuroscientist and sleep researcher, explains, since adenosine continues to build up, you've just been tricked by caffeine, which is hiding its existence from you, and only temporarily.

What I've described here is the direct effect of caffeine on the brain; the chemical also has several indirect effects, including increases in adrenaline, serotonin, and dopamine. The release of dopamine is typical in drugs of abuse, and probably accounts for caffeine's mood-enhancing qualities—the cup of optimism!—as well as the fact that it is habit-forming. Caffeine is also a vasodilator and can be mildly diuretic. It temporarily raises blood pressure and relaxes the body's smooth muscles, which may account for coffee's laxative effect. (This could explain some of coffee's early popularity; constipation was a serious matter in seventeenth- and eighteenth-century Europe.)

But what is unique about caffeine is the targeted way in which it interferes with one of the most important of all biological functions: sleep. Walker, in his 2017 book *Why We Sleep*, argues that the consumption of caffeine—the most widely used psychoactive stimulant in the world—"represents one of the longest and largest unsupervised drug studies ever conducted on the human race." We now know the results of that study and, if Walker is to be believed, they are alarming.

For as long as people have been drinking coffee and tea, medical authorities, as well as quacks of various persuasions, have warned about the perils to human health posed by these beverages, which is to say, the dangers of caffeine. And ever since the seventeenth century, when women worried about coffee's effect on male potency, the presumption has been that there *must* be a problem. Perhaps because we believe more deeply in the iron law of compensation than in the possibility of a free lunch, researchers have undertaken a massive, worldwide, centuries-long search to pinpoint caffeine's karmic payback—the way in which our fond habit must surely be killing us. Cancer? Hypertension? Heart disease? Mental illness? At one time or another, caffeine has been implicated in all these problems and a great many more.

And yet, at least till now, caffeine has been cleared of the most serious charges against it. The current scientific consensus is more than reassuring—in fact, the research suggests that coffee and tea, far from being deleterious to our health, may offer some important benefits, as long as they aren't consumed to excess. Regular coffee consumption is associated with a *decreased* risk of several cancers (including breast, prostate, colorectal, and endometrial), cardiovascular disease, type 2 diabetes, Parkinson's disease, dementia, and possibly depression and suicide. (Though high doses can produce nervousness and anxiety, and chances of committing suicide climb among those who drink eight or more cups a day.)

Coffee and tea are also the leading source of antioxidants in the American diet, a fact that may by itself account for many of the health benefits of coffee and tea. (And you can get these antioxidants by drinking decaf.)* My review of the medical literature on coffee

---

*This may help dissolve an apparent paradox: how can coffee and tea have such a positive effect on health at the same time they are responsible for the poor sleep that negatively affects our health? A 2017 review article found that decaffeinated coffee has much the same beneficial effect on health as caffeinated coffee, suggesting that it may be the antioxidants, rather than the caffeine, that are most important. (Grosso et al., *Annual Review of Nutrition*, 2017.)

and tea made me wonder if my abstention might be compromising not only my mental function but my physical health as well.

However, that was before I read, and then met and interviewed, Matt Walker.

Why We Sleep is one of the scarier books I've read. Walker, an Englishman, is a compact and wired man—I would describe him as caffeinated except that I know he is not. He is single-minded in his mission: to alert the world to an invisible public-health crisis, which is that we are not getting nearly enough sleep, the sleep we are getting stinks, and a principal culprit in this crime against body and mind is caffeine. Caffeine itself might not be bad for you, but the sleep it's stealing from you may have a price: According to Walker, research suggests that insufficient sleep may be a key factor in the development of Alzheimer's disease, arteriosclerosis, stroke, heart failure, depression, anxiety, suicide, and obesity. "The shorter you sleep," he bluntly concludes, "the shorter your lifespan."

Matt Walker grew up in England drinking copious amounts of black tea, morning, noon, and night. He no longer consumes caffeine, save for the small amounts in his occasional cup of decaf. In fact, none of the sleep researchers or experts on circadian rhythms whom I interviewed for this story use caffeine.

I thought of myself as a pretty good sleeper before I met Matt Walker. At lunch he probed me about my sleep habits. I told him I usually get a solid seven hours, fall asleep easily, dream most nights.

"How many times a night do you wake up?" he asked. I'm up three or four times a night (usually to pee), but I almost always fall right back to sleep.

He nodded gravely. "That's really not good, all those interruptions. Sleep quality is just as important as sleep quantity." The interruptions were undermining the amount of "deep," or "slow wave," sleep I was getting, something above and beyond the REM sleep I had always thought was the measure of a good night's shut-eye. But it seems that deep sleep is just as important to our health, and the amount we get tends to decline with age.

During deep sleep, low-frequency brain waves set out from the frontal cortex and travel toward the back of the brain, in the process synchronizing many thousands of brain cells into a kind of neural symphony. This harmonizing of our neurons helps us distill and consolidate the blizzard of information we've taken in during the day. Memories are carried on these slow waves from sites of short-term daily storage to more permanent locations. Picture the mental desktop being cleared off and reorganized at the end of the workday, as the brain's files are stowed in their proper place or trashed.

With all the interruptions I was experiencing, Walker guessed I was sorely deficient in deep sleep. "You probably want to address that." That night he sent me a link for a supplement that purported to improve prostate function.

At the time of our lunch, I hadn't yet begun my abstinence experiment, and Walker inquired about my caffeine use. A cup of half-caf first thing, green tea through the morning, and sometimes, if I'm flagging, a cappuccino after lunch. Walker explained that, for most people, the "quarter life" of caffeine is usually about twelve hours, meaning that 25 percent of the caffeine in a cup of coffee consumed at noon is still circulating in your brain when you go to bed at midnight. That could well be enough to completely wreck your deep sleep.

I shuddered to think about the occasional cup of coffee after dinner. "Some people say they can drink coffee at night and fall right to sleep," Walker said, a note of pity in his voice. "That might be the case, but the amount of slow-wave sleep will drop by fifteen to twenty percent," he said. "For me to decrease your deep sleep by that much, I'd have to age you by twenty percent." Which meant that that after-dinner espresso would give me the lousy night's sleep of a man twelve years my senior. I pictured the anarchy of my computer's desktop after a long day of work when I have neglected to perform any digital hygiene.

Caffeine is not the sole cause of our sleep crisis; screens, alcohol (which is as hard on REM sleep as caffeine is on deep sleep), pharmaceuticals, work schedules, noise and light pollution, and anxiety can all play a role in undermining both the duration and quality of our sleep. But caffeine is at or near the top of the list of culprits. Walker says, "If you plot the rise in the number of Starbucks coffeehouses over the past thirty-five years and the rise in sleep deprivation over that period, the lines look very similar."

(I was relieved to learn that Walker has since eased up a bit on his condemnation of coffee. In a recent exchange, he suggested that the demonstrated health benefits of "moderate morning coffee use" might outweigh the cost to our sleep health. "After all," he wrote, "life is to be lived [to a degree]!")

Here's what's uniquely insidious about caffeine: the drug is not only a leading cause of our sleep deprivation; it is also the principal tool we rely on to remedy the problem. Most of the caffeine consumed today is being used to compensate for the lousy sleep that caffeine causes. Which means that caffeine is helping to hide from our awareness the very problem that caffeine creates. Charles

Czeisler, an expert on sleep and circadian rhythms at Harvard Medical School, put the matter starkly several years ago in a *National Geographic* article by T. R. Reid:

> The principal reason that caffeine is used around the world is to promote wakefulness. But the principal reason that people need that crutch is inadequate sleep. Think about that: We use caffeine to make up for a sleep deficit that is largely the result of using caffeine.

When I recently spoke to Czeisler, he told me he doesn't use caffeine either, but shared a story about his thesis adviser at Stanford, who did. Bill Dement was a legendary sleep researcher, involved in the discovery of the connection between REM sleep and dreams, and the creator of the field of sleep-disorder medicine.

"Once when he stayed with us, he came downstairs in the morning and asked, 'Where's the coffee?' We didn't even own a coffee maker! 'I'm sorry, Bill, but as you well know, caffeine is the enemy of sleep.' 'True,' he replied, 'but it's also the friend of waking!'"

I'm not sure Matt Walker would find that story the least bit funny.

The sleep issue suggests an answer to the conundrum of how caffeine could be a source of human energy. It only looks that way, because caffeine is simply hiding, or postponing, our exhaustion by blocking the action of adenosine. As the liver removes the caffeine from circulation, the dam holding back all that pent-up, still-mounting adenosine will break, and when the rebounding chemical floods the brain you will crash, feeling even more tired than you did before that first cup of coffee. So what will you do then? Probably have another cup.

It appears that there is no free lunch. The energy that cup of coffee or tea has given you has been borrowed, from the future, and must eventually be paid back. What's more, there is interest to be paid on that loan, and it can be calculated in the quantity, and quality, of your sleep.

Our story about the cup of "concentrated sunshine" does seem to be darkening, and I'm afraid it will darken further before it is over. A case can be made that coffee and tea did make a substantial, positive contribution to the advance of quote-unquote "civilization" in the West, if by that we mean the various blessings of culture and capitalism, including the arts and sciences and the standard of living. But just as consumers of caffeine eventually must pay a biological price for the energy supplied by their drug of choice, an economic and even moral price has been paid as well. Almost from the start, the blessings of coffee and tea in the West were inextricably bound up with the sins of slavery and imperialism, in a global system of production organized with such brutal rationality that it could only have been fueled by—*what else?*—caffeine itself.

Coffee and tea, as commodities produced in the global South to be consumed in the North, entangled all who drank them in an intricate new web of international economic relations—specifically colonialism and imperialism. The spice trade—another vibrant market in plant stimulants—preceded the caffeine trade by a few centuries, but it was minuscule by comparison and, on the consuming end, mainly involved the affluent.

By the end of the eighteenth century, tea was being consumed daily by just about everyone in England; it became the most important commodity traded by the British East India Company, accounting for an estimated 5 percent of the nation's gross national product. "It appears a very strange thing," David Davies, an English cleric, observed in the late 1700s, "that the common people of any European nation should be obliged to use, as part of their daily diet, two articles imported from opposite sides of the earth."

The two articles Davies had in mind were tea and sugar, which became paired in England soon after tea's introduction—somewhat surprisingly, since tea in China was never sweetened. No one knows exactly why the practice took root, but the tea imported by Great Britain tended to be bitter and, as a hot beverage, could readily absorb large amounts of sugar. In fact, one of the principal uses of sugar in Britain was as a sweetener of tea, and the custom drove a substantial increase in sugar consumption—which in turn drove an expansion of slavery to run the sugar plantations of the Caribbean. (An estimated 70 percent of the slave trade supported sugar production.) Coffee was even more directly implicated in the institution of slavery, especially in Brazil, where coffee growers imported large numbers of slaves from Africa to work on their plantations. How many tea and coffee drinkers in Europe had any idea that their sober and civilized habit rested on the back of such brutality?

The British East India Company's tea trade with China bore a moral stain of another kind. Since the company had to pay for tea in sterling, and China had little interest in English goods, England began running a ruinous trade deficit with China. The East India Company came up with two clever strategies to improve its balance-of-payments position: It turned to India, a country it controlled that had no history of large-scale tea production, and transformed it into a leading producer of tea—and opium. The tea was exported to England and the opium, over the strenuous objections of the Chinese government, was smuggled into China, in what would quickly become a ruinous and unconscionable flood.

By 1828 the opium trade represented 16 percent of the company's revenues, and within five years, the East India Company was sending more than five million pounds of Indian opium to China per year. This certainly helped close the trade deficit, but millions of Chinese became addicted, contributing to the decline of what had been a great civilization. After the Chinese emperor ordered the seizure of all stores of opium in 1839, Britain declared war to keep the opium flowing. Owing to the Royal Navy's vastly superior firepower, the British quickly prevailed, forcing open five "treaty ports" and taking possession of Hong Kong, in a crushing blow to China's sovereignty and economy.

So here was another moral cost of caffeine: in order for the English mind to be sharpened with tea, the Chinese mind had to be clouded with opium.

Those of us who enjoy a cup of coffee or tea today know scarcely more about the system that produces it than consumers did during the time of slavery or the Opium Wars. The intricate supply chain that delivers us our daily dose of caffeine is largely invisible, and while it no longer rests on the backs of African slaves or Chinese opium addicts, a regime of economic exploitation remains at its base. For every four-dollar latte, only a few pennies ever reach the farmers who grew the beans, most of whom are smallholders working a few steeply raked acres in some rural corner of a tropical country. In recent years, the global price for coffee beans has moved in giant, destructive swings, as the market does what markets do: scours the world for the lowest-price producer at any given moment.

In the 1960s, the world's coffee-growing nations banded together to limit those swings by managing supply cooperatively. The International Coffee Agreement set export quotas for each coffee-producing nation, as a way to keep prices stable within a certain range. This worked for many years. But in 1989, after the rise of neoliberal economics and the consolidation of buying power in the hands of a small number of multinational corporations, the coffee agreement fell apart. Prices now are set by futures markets in London and New York, and move up and down dramatically and unpredictably. In many years, farmers are forced to sell their beans for less than it cost to grow them. Of the ten dollars you may pay for a pound of coffee, only about one dollar reaches the farmer who grew it. At the higher end of the market, a handful of companies like Starbucks and certification schemes like Fairtrade International are seeking to improve the lot of coffee farmers by paying them a guaranteed price. But a free market in any commodity crop that is grown by millions of small producers and bought by only a tiny handful of large buyers will inevitably enrich the latter while tending to impoverish the former.

Perhaps you think I'm painting such a dark picture of coffee and tea because, like those Confederate troops, I've been demoralized by the fact that I can't have any. You also may be wondering why I seem to be reducing the rich, complex culture surrounding these two beverages to brain chemistry and economics. Surely this is an overly reductive way to look at things as wonderful as coffee and tea.

You have a point. I don't mean to take anything away from the intricate cultures that surround tea and coffee and transcend the chemical they share. The epitome of caffeine culture is, of course, the Japanese tea ceremony, which elevates the preparation and consumption of tea to a spiritual practice. With the ceremony's multiple layers of ritual, Zen philosophy, elaborate manners, scripted dialogue, and cherished paraphernalia, one could easily lose sight of the reality that one is consuming a drug.

Why is there no comparable coffee ceremony? (The nearest approximation is the traditional coffee ceremony in Ethiopia, where green coffee beans are roasted over an open flame, ground, and then brewed in a special vessel.) What I find curious is just how different in character and symbolism these two caffeine-delivery systems have become. How did the culture of tea become so much more refined than the brawny culture of coffee? Perhaps it has something to do with the fact that a cup of coffee delivers a stronger jolt than tea, which contains less than half as much caffeine. But drink a second cup of tea and you'll be equally caffeinated, so that can't be the whole story. Perhaps it is the taste, or chemistry, or the region of origin, that explains it, or perhaps the different cultural associations of coffee and tea are simply accidents of the beverages' different histories.

Whatever the reason, the differences are striking. In *The World of Caffeine*, Bennett Alan Weinberg and Bonnie K. Bealer neatly contrast the rival cultures by proposing a series of sharp dualities. These are so obvious that I don't need to tell you which term applies to which beverage:

male/female

boisterous/decorous

bohemian/conventional

obvious/subtle

indulgence/temperance

vice/virtue

passion/spirituality

casual/ceremonial

down-to-earth/elevated

American/English

the frontier/the drawing room

excitement/tranquility

demimonde/society

extroverted/introverted

full-blooded/effete

Occidental/Oriental

work/contemplation

tension/relaxation

spontaneity/deliberation

Beethoven/Mozart

Balzac/Proust

And so on. The various delivery systems for alcohol exhibit a similar degree of elaboration—just think of the cultural signifiers that go with wine versus those belonging to beer or hard liquor.

We humans apparently have a deep desire to complicate things, to embroider the most basic biological response with the rich colors and textures of culture. In fact, the very idea that these drinks each constitute a "delivery system" for a psychoactive compound offends us a bit. But someone who hears the elaborate descriptors for wine without ever having drunk it would have no idea that a key point about this beverage is that it changes consciousness. The same is true for coffee and tea—and certainly *not* true for most of the other liquids we consume. Does anyone think this deeply—this metaphorically—about the psychosensory qualities of orange juice, or milk?

No, tea and coffee are special in this regard. Consider this list of descriptors used for "cupping," or tasting, coffee I stumbled on online. It was compiled by Counter Culture Coffee.

The Vegetal/Earthy/Herb category alone is subdivided into twenty flavor profiles, including leafy greens, hay/straw, tobacco, cedar, fresh wood, and soil. There's Savory, which includes meatlike and leathery. There's Grain and Cereal, subdivided into fresh bread, barley, wheat, rye, graham cracker, granola, and pastry. Sweet and Sugary includes brown sugar, maple syrup, molasses, and cola. The other categories—Nut, Chocolate, Dried Fruit, Berry, Stone Fruit, Citrus, Floral, Spice, and Roast—are also broken down into specific flavors. This list doesn't even include another set of descriptors pertaining to body or "mouth feel," such as tea-like, silky, round, velvety, big, and chewy, or a separate list for undesirable qualities, including mold, fruit decomposition, stale bread, Band-Aid, cardboard, compost, animal hide, and funk/garbage.

How wonderful to be able to discern and name such a panoply of flavors, aromas, and textures—seemingly all of nature—in a cup of coffee! Much the same thing can be said for tea, which has its own evocative sensory vocabulary, positive and negative and purely descriptive. So a particular tea can be faulted for being brassy, bakey, chesty (i.e., exuding the smell of the wooden crate it came in), grassy, tarry, or muddy, or praised for being brisk, bright, biscuit-y, malty, nutty, smoky, or muscat-like. Tasters liken the aroma of tea to flowers (lilac, jasmine, magnolia, osmanthus, orchid, lily, lotus, camellia, lily of the valley); to fruit (lychee, pineapple, coconut, passion fruit, custard apple); and to woods, usually Oriental (aloe, sandalwood, cinnamon tree, young camphor, old camphor). Some of these qualities are purely imaginary, no doubt, but most correspond to one of the hundreds of different molecules found in tea and coffee—the esters, terpenes, amines, acids, ketones, lactones, pyrazines, pyridines, phenols, furans, thiophenes, and thiols that together make up our sensory experience of these beverages.

These flavor and aroma molecules are present in your cup, but how much would that matter if not for that one other molecule, 1,3,7-trimethylxanthine? Would people have ever discovered coffee or tea, let alone continued to drink them for hundreds of years, if not for caffeine? There are countless other seeds and leaves that can be steeped in hot water to make a beverage, and some number of them surely taste better than coffee or tea, but where are the shrines to *those* plants in our homes and offices and shops?

Let's face it: The rococo structures of meaning we've erected atop those psychoactive molecules are just culture's way of dressing up our desire to change consciousness in the finery of metaphor and association. Indeed, what really commends these beverages to us is their association not with wood smoke or stone fruit or biscuits, but with the experience of well-being—of euphoria—they reliably give us.

It is this experience, known to drug researchers as reinforcement, that practically guarantees we will return to tea or coffee or wine. It also has the power to alter our perception of their flavors.

"People are badly deceived when it comes to taste," Roland Griffiths, the Johns Hopkins drug researcher, explained. "It's like saying 'I like the taste of Scotch.' No! This is an acquired, conditioned taste preference. When you pair a taste with a reinforcer like alcohol or caffeine, you will confer a specific preference for that taste."

Caffeine is naturally present in coffee and tea, but typically is added to sodas—so why would soda makers do that? Especially in a beverage marketed to children? The industry has claimed (to the FDA and other regulators) that the caffeine is there as a flavoring, and that they add it for the note of bitterness the alkaloid provides. They actually say this with a straight face. In 2000 Griffiths's lab easily undermined the claim with a double-blind taste test in which cola drinkers were asked to detect differences in colas, some caffeinated and some uncaffeinated. Most couldn't taste the difference. And yet the six top-selling soda brands in the U.S. all contain caffeine (typically about as much as in a cup of tea). Griffiths says that if you pair caffeine with *any* flavor, people will express a preference for that flavor. "Just like when I say 'I love the way Scotch tastes.'"

Griffiths's experiment reminded me of another taste test I'd heard about, but it took me a moment to pinpoint what it was (no doubt because I was still off caffeine): *Geraldine Wright's bees!* Wright had done much the same test on her honeybees, and discovered that they, too, developed a preference for nectar that had been caffeinated. We humans are more like the bees than I realized, just as easily duped, in this case by the soda companies rather than the plants, into preferring whichever brand of sugar water has had caffeine added to it.

The soda makers have figured out what the plants learned to do a long time ago.

The time had come to wrap up my experiment in caffeine deprivation. I had learned what I could from it, had harvested a number of excellent nights' sleep, and was eager to see what a body that had been innocent of caffeine for three months would experience when subjected to a couple of shots of espresso. I had effectively returned myself to the condition of caffeine virgin, and was more than ready to sacrifice that status in order to rejoin the human community of the caffeinated.

I had thought long and hard, even lovingly, about where I'd go to enjoy my first cup. It was definitely going to be coffee; as much as I love tea, I didn't think it could give me the psychoactive jolt I was looking forward to. At first I considered getting my first cup from the Peet's in my neighborhood, which happens to be the original Peet's, founded in 1966. On the corner of Walnut and Vine in North Berkeley, Peet's is now something of a landmark, the site of a watershed moment in coffee history. It was Alfred Peet, the émigré son of a Dutch coffee roaster, who almost single-handedly introduced America to good coffee. Before Peet opened his shop, Americans mostly drank instant or diner coffee from blue-and-white cardboard cups or percolated coffee made from cans of Folgers or Maxwell House grounds. At the time, most of this coffee was made from inferior Robusta beans, which are high in caffeine but bitter and one-dimensional in taste. But it was cheap and it was all we knew.

Peet, who had tasted better in the Netherlands, insisted on

sourcing Arabica beans exclusively, and roasting them slowly, until they were quite dark. His exacting standards and Old World aesthetic did much to create the coffee culture in which we now live. A generous man, Peet mentored a whole generation of American coffee importers and roasters, including the founders of Starbucks, who worked for him at the Berkeley shop, learning how to select beans and roast them. Peet also taught Americans to pay a few dollars, rather than a quarter or two, for a cup of coffee, transforming it into a new kind of everyday luxury good. So there would be a certain poetic logic to having my first cup at this local shrine to good coffee.

Alas, I don't *love* Peet's coffee. Too often it tastes burnt. So in the end I decided to honor a more personal coffee tradition. I would opt for a "special" at the Cheese Board, the shop down on Shattuck Avenue where Judith and I have been morning regulars for many years. A special is the Cheese Board's term for a double-shot espresso drink made with somewhat less steamed milk than the typical cappuccino; I believe it's what the Australians would call a flat white.

Out in front of the Cheese Board, a couple of parking spaces have been converted into a sweet little pocket park, with a few benches, a couple of flower planters and trees, and a thick wooden counter to lean on. I seldom take the time to hang out there, but this was such a lovely midsummer Saturday morning that we decided to linger, finding a seat where we could enjoy our coffees and take in the scene. It was still early, so there were lots of paper-cup-toting young parents with little kids deeply absorbed in their muffins and chocolate chip scones. The kids were having their own drug experience.

My special was *unbelievably* good, a ringing reminder of what a poor counterfeit decaf is; here were whole dimensions and depths of flavor that I had completely forgotten about! I could almost feel the tiny molecules of caffeine spreading through my body, fanning out along the arterial pathways, sliding effortlessly through the walls of my cells, slipping across the blood-brain barrier to take up stations in my adenosine receptors. "Well-being" was the term that best described the first feeling I registered, and this built and spread and coalesced until I decided "euphoria" was warranted. And yet there was none of the perceptual distortion that I associate with most other psychoactive drugs; my consciousness felt perfectly transparent, as if I were intoxicated on sobriety.

But this was not the familiar caffeine feeling—the happy (and grateful) return to baseline, as the first cup disperses the gathering fogs of withdrawal. No, this was something well up from baseline, almost as if my cup had been spiked with something stronger, something like cocaine or speed. *Wow—this stuff is legal?* I looked around me, taking in the mellow sidewalk scene, the kids in their strollers, and the dogs trailing them for crumbs. Everything in my visual field seemed pleasantly italicized, filmic, and I wondered if all these people with their cardboard-sleeve-swaddled cups had any idea what a powerful drug they were sipping. But how could they? They had long ago become habituated to caffeine, and were now using it for another purpose entirely. Baseline maintenance, that is, plus a welcome little lift. I felt lucky that this more powerful experience was available to me. This—along with the stellar sleeps—was the wonderful dividend of my investment in abstention.

And yet in a few days' time I would be them, caffeine-tolerant and addicted all over again. I wondered: Was there any way to preserve the power of this drug? Could I devise a new relationship to caffeine? Maybe treat it more like a psychedelic—say, something to be taken only on occasion, and with a greater degree of ceremony and intention. Maybe drink coffee just on Saturdays? I resolved to try.

After about a half an hour, I could feel the initial surge of optimism morph into something a bit more manic and tetchy. A garbage truck had pulled up to the curb outside a restaurant across the street. Unignorable, it began violently shaking tall plastic bins into its maw and then noisily devouring the garbage. The racket was unbearable—or so it felt, in what was becoming, I realized, a hypervigilant state. I began to feel antsy, and started composing lists in my head of things I needed to get done that day. I asked Judith if she was ready to go and she agreed—the scene had lost its charm. So we walked back up the hill and home.

Judith left to go to her studio, and I was left to do, well, whatever I wanted to do—while away the Saturday morning, putter in the garden, maybe make a few calls. But the caffeine had another idea. It wanted me to tackle my to-do list, harness the surge of energy—*of focus!*—coursing through me, and put it to some good use.

For some reason this had everything to do with throwing stuff out. I went to my computer and systematically "unsubscribed" from at least a hundred Listservs that had been clogging my in-box. This felt *great*. Until I felt too antsy to sit at my desk a moment longer. Another task suddenly demanded my attention: it was time to tackle my closet! This is not something I have *ever* done of my own free will, but at that moment I wanted nothing more than to take all my sweaters off the shelf and sort them into four piles: in need of laundering, moth-eaten discards, giveaways, still in rotation. Ordinarily, I feel faithful to my old clothes and have a hard time accepting that any item has outlived its usefulness. But not today. Today I was merciless, and quickly filled a large garbage bag with not only sweaters but also sneakers, shirts, even sport jackets, all destined for Goodwill and good riddance.

The morning went on like that, as I compulsively got stuff done—on the computer, in my closet, in the garden and the shed. I raked, I weeded, I put things in order, as if I were possessed, as I guess I was. Whatever I focused on, I focused on zealously and single-mindedly. I was like a horse wearing blinders; the periphery and its distractions had completely vanished from my field of awareness. I could sink myself into a task and easily fail to notice that an hour had passed.

Around noon, my compulsiveness began to subside, and I felt ready for a change of scene. I had yanked a few plants out of the vegetable garden that were not pulling their weight, and decided to go to the garden center to buy some replacements. It was during the drive down Solano Avenue that I began idly fantasizing about how I might get a second cup, and all at once I realized the true reason I was heading to this particular garden center: Flowerland had this Airstream trailer parked out front that served really good espresso drinks.

I had only had a single cup of coffee after three months of abstinence, and already the insidious tentacles of dependence were wrapping themselves around me! What had happened to my hours-old resolution to drink coffee only on Saturday? Then I heard a voice say,

But it is *still Saturday!* I knew immediately who it was: the clever and sinuous voice of the addict. It took all the willpower I could muster to resist.

Partway through my research for this story, it occurred to me that I had never actually laid eyes on a coffee or tea plant. Well, that's not *entirely* true: a few years ago, the Peet's in my neighborhood kept a rather sad and scraggly coffee plant in a pot by the door, but it never bore fruit and didn't survive long. I had certainly never encountered a coffee plant in its native habitat. So I decided to pay *Coffea arabica* a visit.

Other business had brought us to Medellín, the gateway to Colombia's premiere coffee-growing region, so, on a January morning, Judith and I hired a car to take us up into the mountains south of the city. Our destination was Café de la Cima, a coffee farm, or finca, a few miles by rutted dirt road outside of Fredonia, a lively little market town stretched out in the shadow of Cerro Bravo. Along the way, we passed Cerro Tusa, the perfect green triangle of a volcano depicted on the logo for Colombian coffee. You've seen it a thousand times on packages of beans and in all those commercials for Colombian coffee—the classic ones featuring Juan Valdez.

It turns out Juan Valdez is a purely fictional campesino. He was conceived in the brain of an advertising copywriter in the Manhattan offices of Doyle Dane Bernbach, the ad agency, in 1958, for the purpose of selling Colombian coffee to the world. Octavio Acevedo and his son Humberto, the proprietors of Café de la Cima, could have served as role models for Valdez, right down to the straw hat and colorful serape. (The only thing missing from the scene is

Conchita, Valdez's faithful burro.) Humberto, who showed us around the seven-acre finca, is the fourth generation to grow coffee on this steep, lush hillside. But the operation has changed in important ways since his grandfather farmed it.

"Five years ago," Humberto explained as we set out to visit his plants, "my father decided he wanted to taste the coffee he was growing." This was a radical idea; most campesinos sell their beans to middlemen while they're still "green"—freshly picked and unprocessed. If they drink coffee at all, it's coffee grown by someone else and is probably *tinto*—the thick, concentrated coffee made from cheap beans that most Colombians still drink. All the best beans go to the export market. But Octavio could see there was no future for a small farmer selling a commodity crop into what has become a turbulent global market, so he decided that he would try to sell something different: coffee that had been grown, harvested, cleaned, fermented, dried, and roasted on the farm. Café de la Cima would become a brand in an artisanal coffee market, as well as a destination for people like me, curious to see where and how their coffee is produced.

Humberto was eager to introduce us to the twelve thousand coffee plants, a mix of Bourbon and Castillo, with whom the family shares this verdant, sun-drenched hillside. Coffee likes tropical mountains because the plant needs both plenty of rain and exceptionally good drainage in order to thrive. Growing at higher elevations—Café de la Cima is perched sixteen hundred meters above sea level—also allows coffee to escape one of its most destructive pests, the fungus that causes coffee leaf rust.

Climate change is already pushing coffee production higher up the mountain and making life difficult for farmers. Coffee plants are notoriously picky about rainfall, temperature, and sunlight, all of which are changing in Colombia, rendering lands that had always been good for coffee production no longer viable. Worldwide, the prospects for coffee production in a changing climate are, according to the agronomists, dismal. By one estimate, roughly half the world's coffee-growing acreage—and an even greater proportion in Latin America—will be unable to support the plant by 2050, making coffee one of the crops most immediately endangered by climate change. Capitalism, having benefited enormously from its symbiotic relationship with coffee, now threatens to kill the golden goose.

Humberto led us up a steep path behind the house. We passed a nursery where he was sprouting coffee plants—dozens of tiny seedlings, each wearing a tan coffee bean like a cleft hat. It's easy to forget that coffee beans are first and foremost seeds. Rather than buy replacement plants when their production declines, Humberto has begun selecting and germinating his own, scouring the farm for specimens that thrive in his particular soil and microclimate.

Up past the nursery, we crossed a little stream and stepped into the first row of coffee plants: curving parallel lines of five-foot-tall pruned shrubs thick with glossy green leaves and bearing slender branches lined with "cherries." Most of the fruits were still green, but there were a handful of bright red ones that looked more like cranberries than cherries. Humberto handed Judith and me each a basket, worn in front at waist height and suspended by a strap over the shoulder. He shooed us away: *Go pick some coffee!*

We each went our own way, stepping gingerly down a different narrow row of spiky green shrubs. The hillside is so steep I had to carefully sidestep my way from plant to plant, bending over and reaching through the leaves to pick only the reddest cherries, one by one, and dropping them into my basket. I bit into a ripe red one. The flesh tasted fruity and sweet, with just a suggestion of coffee flavor, and in the center sat a small tan seed, divided into two lobes like a miniature pair of buttocks.

Humberto had told me it takes fifty or so coffee beans to make a single cup of coffee; after a half hour of picking, I had collected enough beans for maybe four or five cups, and already my back and feet were in an uproar of pain. It was hard to believe coffee was still picked by hand, bean by bean—that so little has changed over the centuries. But the steepness of the orchards discourages both mechanization and consolidation; this is still an agriculture dominated by millions of small farmers with readier access to hands than to capital.

The biggest innovation at Café de la Cima is the one that has put Conchita out of work. When a picker's basket is topped up, he or she no longer straps it to the back of a burro for the ride down the hillside. Now the picker spills the basket of coffee cherries into a concrete box at the top of the hill; a stream of well water then flushes the cherries through a steel pipe, carrying them down the mountain and directly into the processing shed.

I didn't pick enough coffee to fill my basket, not even close. I was stymied by having to stretch my legs and straighten up every few minutes, or my back would bitterly complain. The hillside was so precipitous and the rows so tightly planted that I found it difficult to secure a confident purchase with my feet. I felt off-balance the entire time, which made it hard to work efficiently. Among the coffee shrubs, I felt like an interloper, a stranger in a habitat far more congenial to them than to a biped.

As I stepped out of the row I'd been working and gazed out over the Andes, one verdant fold overlapping another, I could see rows of shiny green coffee plants snaking across the landscape, each following the horizontal contours and stepping up the sheer flanks of the mountains. It was hard to imagine how this remote and sleepy rural scene had anything whatsoever to do with our everyday urban lives, but one doesn't exist without the other. The two realms have become intimately connected, and are now implicated in each other's destinies by powerful vectors of trade and desire. Our taste for coffee, only a few hundred years old, has reconfigured not only this landscape and the lives of the people who tend it, but the very rhythms of our civilization.

Yet it wasn't coffee's taste alone that worked those wonders; it was also, crucially, the tiny molecule that contributed the bitterness to that drink, and what that molecule did to our minds once it found its way into our brains. What is impossible to see from this distance is how all the glossy green leaves blanketing these mountains are at this very moment transforming the strong rays of the tropical sun and the nutrients in these ruddy soils into 1,3,7-trimethylxanthine. The plants have turned these mountainsides into factories for the production of caffeine. What is difficult to square, standing here, is how a landscape as unhurried and tranquil as this one could be the driver of so much speed, energy, and industry in the world I'll soon return to.

Perched somewhat crookedly on the steep slope of one of these caffeine mountains, my main thought was, *You really have to give this plant a lot of credit*. In less than a thousand years it has managed to get itself from its evolutionary birthplace in Ethiopia all the way here to the mountains of South America and beyond, using our species as its vector. Consider all we've done on this plant's behalf:

allotted it more than 27 million acres of new habitat, assigned 25 million humans to carefully tend it, and bid up its price until it became one of the most precious crops on earth.

This astounding success is owing to one of the cleverest evolutionary strategies ever chanced upon by a plant: the trick of producing a psychoactive compound that happens to fire the minds of one especially clever primate, inspiring that animal to heroic feats of industriousness, many of which ultimately redound to the benefit of the plant itself. For coffee and tea have not only benefited by gratifying human desire, as have so many other plants, but these two have also assisted in the construction of precisely the kind of civilization in which they could best thrive: a world ringed by global trade, driven by consumer capitalism, and dominated by a species that by now can barely get out of bed without their help.

Of course, this all began strictly as an accident of history and biology—remember the goats that were said to have inspired that curious herder to taste his first coffee berry? But that's how evolution works: nature's most propitious accidents become evolutionary strategies for world domination. Who could have guessed that a secondary metabolite produced by plants to poison insects would also deliver an energizing bolt of pleasure to a human brain, and then turn out to alter that brain's neurochemistry in a way that made those plants indispensable?

The question arises: which party is getting the better of the symbiotic arrangement between *Homo sapiens* and these two great caffeine-producing plants? We probably lack the perspective needed to judge the question impartially, or to perceive how a plant we "use" might actually be using us. Big-brained and self-regarding primates that we are, we automatically assume we've been calling the shots with these two "domesticated" species, transporting and planting them where we choose, earning billions off them, and deploying them to gratify our needs and desires. *We're in charge*, we tell ourselves. But isn't that exactly what you would expect an addict to say? *Sure you are*. Bear in mind that caffeine has been known to produce delusions of power in the humans who consume it, and that this story of world-conquering success would read very differently had the plants themselves been able to write it.

My own relationship to caffeine remains a work in progress. I've been trying to honor the epiphany I had during my coffee "trip" (which is how I remember it)—that there is a better way to relate to coffee than as an addict, one that would safeguard both my agency and the plant's power. So for several weeks I drank caffeinated coffee only on Saturdays. This so dramatically improved my Saturdays that I gradually found myself slipping in a little caffeine during the week—maybe a cup of green tea to clear out the dregs of a particularly muzzy morning, or a decaf when I wanted to treat myself to the taste of coffee. But as with so many addictions, the slope is slippery; the mind concocts elaborate arguments for the purpose of undermining its best intentions. I suspect it's easier to enforce an absolute ban than one that's shot through with exceptions and therefore subject to rationalization and self-deception.

My latest idea is simple: to have some caffeine on Saturdays, for pleasure (and household chores), but also at a few select other times, "when I need it." Use coffee, or tea, as a tool, in other words, rather than let coffee and tea use me. I remember Roland Griffiths telling me that there had been a time in his life when he used caffeine in precisely this way—only when he had a big deadline, say, or was writing a grant. True, when he told me this he was sipping a tall Starbucks, suggesting either that Skyping with me qualified as a special occasion or that the regime had eventually crumbled. But maybe I could sustain it. I would try.

Take this morning as an example. It is not Saturday. I'm writing the last paragraphs of this story, always a fraught exercise. People talk a lot about the importance of beginnings, but endings matter, too—ideally, they should strike a bell that reverberates long after you've closed the book. (Assuming you've gotten this far—but you have.) I've put off writing this ending several days in a row, not sure exactly how to handle it. You'll recall I began this story in a bit of a crisis, having given up caffeine (for the story) and with it my confidence in the value of what I was about to write. Eventually I found my bearings, though, and managed to rekindle interest in the subject without the use of the subject. I had broken free of caffeine's grip, or so I liked to think.

This morning, however, setting out to find these last words, up against my deadline, I felt like I needed, and, honestly, I deserved, a little something extra to push me over the finish line. But it was only Thursday. Was this a strong enough reason to break my Saturday rule? As Judith and I walked down the hill to the Cheese Board this morning, I was unsure what I would order right up to the moment when I stepped to the front of the line. It was not just the barista I surprised when these words popped out of my mouth:

"Make it a regular, please."

MESCALINE

1. *The Door in the Wall*

It was all set, everything coming together perfectly: the reporting trips scheduled, the access nailed down—all the narrative elements of my mescaline story were falling neatly into place. In April I would fly to Laredo and drive out to the Peyote Gardens, the strip of thorn-scrub running along either side of the Rio Grande, and the only place in the world where the peyote cactus grows wild. A cactusologist (cactologist?—not sure) named Martin Terry had offered to give me a tour, after which we would meet up with a group of Native Americans from several tribes on their annual pilgrimage to gather the inconspicuous little cacti for their ceremonies. In Western culture, peyote is a relatively obscure "psychedelic," but it's a precious sacrament in the Native American Church, the pan-tribal religion that sprang up in the 1880s, at the moment when Indian civilization in North America stood on the verge of annihilation. Native Americans I had interviewed claimed that their peyote ceremonies had done more to heal the wounds of genocide, colonialism, and alcoholism than anything else they had tried. I had arranged an opportunity to see for myself: an invitation to observe and, with luck, take part in a peyote meeting, a meticulously choreographed all-night ceremony typically conducted around a fire in a tepee. And then there was the whole San Pedro angle—San Pedro being the other mescaline-producing cactus, this one from the Andes, where it has been used by Indigenous peoples for centuries before the Spanish conquest. A shaman from Cuzco named Don Victor was coming to Berkeley to lead a ceremony to which I had wangled an invitation. The mescaline piece was starting to write itself. I was excited: this was a story that promised to take me some distance from my accustomed world—not only geographically but culturally, pharmacologically (I had never tried mescaline in any form), even linguistically, since I was venturing into a realm where Western terms I relied on, like "drug" and "psychedelic," were considered offensive. I'd heard about a journalist who, speaking to a Huichol shaman, had referred to peyote as a drug. "Aspirin is a drug," the shaman replied. "Peyote is sacred."

And then, in mid-March, the pandemic burst upon the world, upending all our plans. Don Victor couldn't travel. The pilgrimage to Texas was called off and the November ceremony put on hold. Maybe things would be better by November—everyone involved hoped so—but as spring turned into summer and the virus failed to loosen its grip, I began to lose hope that I could travel or do any reporting that wasn't confined to Zoom. The whole idea of travel, of expanding one's knowledge—*one's mind!*—with new sights and experiences, had suddenly become unthinkable. It felt as though one's mental horizon had suddenly and dramatically been foreshortened, that the possibilities of experience, at least those that depend on movement and human contact, were contracting. For how long, nobody knew.

Not that this was *all* bad; 2020 brought the most beautiful spring anyone could remember, mainly, I suspect, because it was the first spring any of us had slowed down long enough to fully notice. Judith and I were walking the Berkeley Hills every morning and evening, charting, week by week, the unfurling of the floral calendar: March's magnolias and camellias giving way to April's wisteria, May's fragrant jasmine and roses to June's poppies and daisies. Nature went gloriously on, oblivious to the virus.

But after several borderline blissful weeks of what we began to refer to as "the pause," a low-grade claustrophobia began to set in. When Fauci* said we could expect another year of this, I was forced to face the fact that "this" was life now, and for as long as we could see. The novel experiences I had put on hold were probably *never* going to happen. The life chapter I was looking forward to writing—the chapter about mescaline and what it had to teach me, about everything from Indigenous culture to the birth of a new religion, from the botany of cacti to the possibilities of human consciousness—was probably going to remain blank, canceled, like so much else, by COVID.

After a few days of feeling unreasonably sorry for myself—for on the scale of 2020's losses, mine were weightless—I decided I should try to think about the problem a little differently. Sure, I could wait for the vaccine, call my editor, and put the story off for a year or year plus. Or I could choose to regard this obstruction that history and life had placed in my way as a spur to think harder or more inventively, as something to be surmounted or circumnavigated or somehow passed through. Somehow.

And then one sun-drenched June afternoon, as spring 2020 made

*For future readers, "Fauci" is the man everyone in America once knew as Dr. Anthony Fauci, the director of the National Institute of Allergy and Infectious Diseases and an adviser to the U.S. government on the novel coronavirus. At the time of this writing, he needed no introduction or first name.

the turn toward the first summer of the pandemic, I found myself rereading *The Doors of Perception*, Aldous Huxley's classic account of his first experience with mescaline in 1953. Huxley describes a "principal appetite of the soul" for a means of transcending the limitations of circumstance, the various walls—whether of habit or convention or selfhood—that confine us. For him, it was mescaline that had shown him a "door in the wall."

That's when it dawned on me: maybe mescaline itself might hold an answer, might point the way around, or through, the obstacle I was confronting. If ever there was a story that should be tellable without physically leaving home, surely it was one about a molecule that transported the mind to new places, the kind of places that couldn't be locked down. Mind you, I say this as someone who had never tried mescaline, whether in the form of peyote, San Pedro, or a synthetic crystal in a pill, and I hadn't a clue how to procure some. But this hopeful if possibly crackpot idea had taken root: maybe mescaline was not merely the subject of the story, but also, somehow, the tool that would allow me to tell it without going anywhere. Along, that is, with Zoom.

2. *The Orphan Psychedelic*

My fascination with mescaline is a fairly recent development. When I first read Huxley, in the 1990s, I hadn't yet tried any of the "classic" psychedelics, so I tended to lump them all together, and read the book as an account of the sort of experience that any psychedelic could sponsor. In 1954, when *The Doors of Perception* was published, LSD had only recently been introduced (by Sandoz Laboratories in the late 1940s), and it would be another few years before the West learned about psilocybin, with the 1957 publication of Gordon Wasson's account of the "mushrooms that cause strange visions" in *Life* magazine. Though the word "psychedelic" wouldn't be coined until 1956, Huxley's account of his 1953 mescaline journey stood, and stands still, as the canonical "psychedelic trip."

It was only after I had sampled the longer menu of psychedelic molecules—LSD, psilocybin, 5-MeO-DMT, and ayahuasca—that I began to wonder about mescaline, which had become a fairly obscure entrée on that menu, rarely encountered and seldom discussed. Now, rereading Huxley after having had those experiences, I could appreciate how distinct mescaline was from the other psychedelics. Huxley didn't describe leaving the known universe, journeying to a "Beyond" populated by strange characters or decorated with extraordinary visual patterns; indeed, he reported no hallucinations. He didn't travel inward to plumb the depths of his psyche or to recover suppressed memories. Nor did his ego dissolve, allowing him to merge with the universe or god or nature. He didn't report the (classic) psychedelic epiphany that love is the most important thing in the universe.

No, Huxley remained very much on this earth, sitting in his Los Angeles garden, observing the familiar physical world—but through completely new eyes:

> "This is how one ought to see," I kept saying as I looked down at my trousers or glanced at the jeweled books in the shelves, at the legs of my infinitely more than Van-Goghian chair. "This is how one ought to see, how things really are."

Huxley suffered from poor eyesight, but not on this particular afternoon. Now the material world revealed itself to him in all its beauty, detail, profundity, and "Suchness"—as it *really* was, whatever that means. (I wonder: does the novelty and power of this sort of radical noticing impress women as much as men? I tend to doubt it.) Huxley spent hours (and pages) dilating on the "is-ness" of a chair, a bouquet of flowers, and the folds of his gray flannel trousers, entranced by "the miraculous fact of sheer existence." These objects weren't getting up and dancing, or transforming themselves into the god Shiva, or talking to him—they were just *being*, and what an astonishment that was!

"How things really are." The question arises: why don't we see this way all the time? Huxley suggests ordinary consciousness evolved to keep this information from us for a good reason: to prevent us from being continuously astonished, so that we might get up from our chair now and again and go about the business of living. Huxley recognized the danger of being constantly thunderstruck by reality: "For if one always saw like this, one would never want to do anything else."

That's why our usual perception of the world is "limited to what is biologically or socially useful"; our brains evolved to admit to our awareness only the "measly trickle" of information required for our survival and no more. Yet there is much more to reality, and 400 milligrams of mescaline sulfate was what it took to throw open what Huxley calls "the reducing valve" of consciousness—aka the doors of perception.

Reading Huxley's account while quarantined in a pandemic intensified my desire to try mescaline. The idea that a molecule could somehow deepen or expand the scope of one's reality suggested a mental strategy nicely tailored to the situation. I was reminded of the lovely line Shakespeare gave Hamlet, enduring a different kind of claustrophobia: "I could be bounded in a nutshell, and count myself a king of infinite space." Mescaline might offer a way to do that, not as a means of escape from circumstance, but as an expansion of it. Instead of an alternate reality, it promised infinitely more of this one.

Huxley experimented with mescaline because he wished to learn something about his mind and its relation to reality. No doubt what he learned was influenced by his mind's own predilections and prior concepts, as much as he claimed he wished to escape them by accessing something nearer to "direct perception" of reality. (If there is a villain in *The Doors of Perception*, it is the constraining power of words and concepts—ironic, perhaps, for a writer, or perhaps not, since writers are acutely aware of the limitations and betrayals of their principal tool.) His specific concerns and motivations—as a Western intellectual and writer, as an Englishman living in Los Angeles, as a "poor visualizer"—all play a role in shaping his experience on mescaline. Huxley may talk about "direct perception," but the man can't look at a chair without thinking about Van Gogh, or at the creases in his trousers without thinking about folded cloth in a Botticelli. Though Huxley does at times make reference to art and thought from the East, the set and setting of his experience could hardly be more Western, or more white.

Yet the molecular hero of Huxley's book came to the West from the Native peoples, and native flora, of North America—call it a gift or, as some might now, a theft. Although it was a German chemist who, in 1897, first isolated the psychoactive molecule in *Lophophora williamsii*, the peyote cactus, and in 1919 an Austrian chemist who first synthesized mescaline, the cactus itself has been used by the Indigenous peoples of North America for at least six thousand years, making it the oldest-known psychedelic, as well as the first to be studied by Science and ingested by curious Westerners.

Some of those curious Westerners were acutely aware of, and specifically attracted to, the Otherness mescaline represented to them. Antonin Artaud, the French author and dramatist (1896–1948), was drawn to mescaline precisely because it "was not made for Whites." He encountered Tarahumara in Mexico, who tried to prevent him from using it because it might offend the spirits. "A white, for these Red men, is one whom the spirits have abandoned." For cosmopolitan Westerners like Artaud, mescaline held the power to re-enchant a world from which the gods had departed.

Though the same chemistry is in play, the uses and meanings of synthetic mescaline for Westerners and the peyote cactus for Indigenous peoples could scarcely be more different. The importance of Timothy Leary's notion of set and setting as shapers of the psychedelic experience surely applies at the level of cultures as well as individuals. The use of the word "chemistry" in the sentence above betrays my own orientation. Yet my hope in exploring the two worlds of mescaline—Western and Indigenous—was to at least try to understand, if not bridge, the gulf that separates them. Did Huxley's account of mescaline (or mine, assuming I got to write one) in any way map the Native American experience of peyote? Did the phenomenology he describes—the almost devotional absorption in the given world—in any way rhyme with the Indigenous understanding of nature not merely as a symbol of spirit but as immanent—a manifestation of it? I was struck by the timing of their embrace of peyote, just when their world was being radically circumscribed—to the tightly bounded dimensions, you might say, of a nutshell. It was in the 1880s, soon after the Plains Indians, kings of infinite space, had lost their freedom to roam the West, and been confined to reservations, that they turned to peyote in order to achieve or recover . . . what exactly?

A more immediate and prosaic question I needed to answer first was: what happened to mescaline in the West after Huxley told everyone how amazing it was? It seemed to have disappeared. At the same time that the use of peyote among Native Americans is growing rapidly (to the point where shortages of the cactus have become an urgent concern), mescaline has become virtually impossible to find. And now, in the midst of a renaissance of scientific research into psychedelics, I hadn't heard of a single research project in the U.S. involving this particular psychedelic.*

I wondered if it was because LSD and psilocybin are simply better drugs, but when I began asking around in the "psychedelic community," invariably I heard precisely the opposite. Everybody loved mescaline! A thirtysomething psychonaut of wide experience told me that when he had finally gotten ahold of some synthetic mescaline recently, he could hardly believe what he'd been missing.

"Why have you been keeping this from us?!" he wondered, referring to his psychedelic elders. "All this time, the hippies were hiding the best drug!" He spoke of the "warmth," "gentleness," and "lucidity" of mescaline, qualities he compared favorably to the

*I've since learned of mescaline research projects in the planning stages, one at the University of Alabama and the other at a psychedelic pharmaceutical start-up in the Bay Area called Journey Colab.

hard-edged "jangliness" of LSD and the more-than-occasional terrors of ayahuasca.

One of those psychedelic elders is a woman in her sixties I spoke to by Zoom. Evelyn, as I'll call her, has been leading a mescaline circle—an all-night ceremony very loosely based on Indigenous peyote rituals—in Northern California since the 1980s. She feels there is something about this particular medicine ("Please let's not call it a drug") that lends itself to the social experience of a ceremony, as well as to the playing and singing of music. (In her ceremony participants sing show tunes.)

"People can stay attuned to one another on mescaline," Evelyn explained. "It doesn't send you to Alpha Centauri, so you're less likely to become an embarrassment to the psyche." Evelyn's description of her ceremony made me realize that the crisp line I was drawing between Western and Indigenous uses of mescaline might blur in places, and that sticky questions of cultural appropriation loomed ahead.

Another psychedelic elder, a rabbi I know with a long-standing interest in psychedelic therapy, was definitive: "Mescaline is the king of the materials." He reminded me that Alexander "Sasha" Shulgin, the legendary psychedelic chemist, shared this assessment. Shulgin, who had worked as a chemist at Dupont before he discovered his vocation in the course of a mescaline trip in the late fifties, synthesized hundreds of new psychedelic compounds, working in his backyard laboratory in Lafayette, California. Many of them involved tweaking the chemical structure of mescaline, which he declared his favorite. (The DEA so respected Shulgin's expertise that they turned to him whenever they seized a drug they couldn't identify; in exchange, they granted Shulgin a DEA license allowing him to work with Schedule I compounds.)

Shulgin's transformative trip took place just a few years after Huxley's: "A day that will remain blazingly vivid in my memory, and one which unquestionably confirmed the entire direction of my life." He describes being able to perceive hundreds of nuances of color that he had never seen before. "More than anything else," he wrote years later, "the world amazed me, in that I saw it as I had when I was a child.

"The most compelling insight of that day was that this awesome recall had been brought about by a fraction of a gram of a white solid, but that in no way whatsoever could it be argued that these memories had been contained within the white solid." Rather, they came from the psyche, he realized, which, whether we realize it or not, contains an "entire universe," and there "are chemicals that can catalyze its availability."*

I asked the rabbi why he thought "the king of materials" had become so scarce. "The thought might arise," he suggested, referring to someone in the midst of a mescaline experience, "*When is this going to end?*" A mescaline trip can last fourteen hours. "It's a commitment," he said. This probably explains its absence from scientific research—psilocybin, the psychedelic typically used in experiments and drug trials, lasts less than half as long, allowing everyone involved to get home in time for dinner. Another strike against mescaline is that a dose requires up to half a gram of the chemical; compare that to LSD, doses of which are measured in micrograms—millionths of a gram. In the illicit drug trade, more material means more risk. Which probably explains why LSD,

---

*Shulgin titled his memoir *PiHKAL: A Chemical Love Story*. The word "PiHKAL" is Shulgin's acronym for "Phenethylamines I Have Known and Loved." Phenethylamines comprise a class of organic compounds found in plants and animals, and that includes both mescaline and MDMA, also known as Ecstasy.

virtually weightless and easy to hide, came to eclipse mescaline, rendering it, by the mid-1960s, an orphan psychedelic.

As for plant sources of mescaline, most of the peyote gathered in Texas ends up in the hands of Native Americans, who have enjoyed the legal right to consume it since President Clinton signed the American Indian Religious Freedom Act Amendments in 1994. I was told it is virtually impossible to come by peyote today if you are not a tribal member. It is also a federal crime for a non-Native person to possess it, grow it, transport it, buy it, sell it, or ingest it. Which, according to many Native Americans, is exactly as it should be. Given the importance of peyote to Native Americans today, and the shortages of the cactus, surely they have a point.

Then there is the San Pedro cactus, which also produces mescaline, albeit at lower concentrations. No, I had never heard of it, either. But it turns out that San Pedro, which is native to the Andes, has become commonplace in California, where it is planted as an ornamental and, unlike the peyote cactus, is perfectly legal to grow. Oddly, however, few Americans or Europeans beyond a tiny community of aficionados seem to know about San Pedro. One of these aficionados told me it grows all over Berkeley, you just need to know what to look for. Could it be that the object of my desire was hidden in plain sight?

3. In Which We Meet the Cacti

So it was: as it happens, not only does San Pedro grow all over Berkeley, but a specimen of the cactus has been happily growing in my own garden for several years now, without the gardener quite knowing it. That's because the person who gave me a cutting of it several years ago didn't call it San Pedro. He called it by its Quechua name, Wachuma.

The son of old friends, Willee had traveled to Peru during a gap year and tumbled into the world of shamanism and plant medicine. He had planted a half dozen or so Wachuma in his parents' backyard, and when we were there for dinner several years ago, he gave me a cutting to take home. Willee explained that Wachuma is a sacred medicine plant in Peru, but at the time I failed to make the connection to mescaline. (Scientists had long failed to make that connection, too: it wasn't until 1960 that mescaline was identified as the psychoactive alkaloid in Wachuma.) I'm always happy to introduce another psychoactive plant to my garden, so was pleased to have it. He also informed me that my cactus was descended from a plant originally propagated from cuttings taken from Sasha Shulgin's garden. My new cactus had a distinguished pedigree.

San Pedro, I learned later, is the Christian name for the Wachuma cactus, named for the saint who held the keys to the gates of heaven. The name at once hinted at the power of the plant and served to mollify the Spanish, who as Catholics had a problem with the idea of an alternative sacrament, and a plant sacrament at that. (The Native American Church made a similar move a few centuries later, when it adopted several Christian elements, such as calling itself a Church, lest the new religion seem too overtly pagan.)

I planted the two-inch cutting in a pot of cactus mix, kept it moist for a few weeks until it rooted, and then, quickly for a cactus, it began to send up a trio of elegant columns of differing heights—a candelabra. The skin was a smooth matte green with a slight bluish tint. The columns (or "candles," as the cactologists say) are divided into six vertical ribs, each punctuated every few inches by an areole from which jut exactly five short, sharp spines. The vertical ribs come together at the top of each column to form a six-pointed star. It's a handsome cactus, stately and architectural, a bit like the model for a Gaudí-esque skyscraper.

I've taken a much more active interest in my cactus since learning it is busy transforming sunlight into mescaline right in my front yard. But how to get from this to that, from the plant to an ingestible psychoactive compound, I had no clue; nor did I know if my cactus was anywhere near ready to harvest.

I reached out to Keeper Trout, one of the world's foremost experts on San Pedro. Alas, it turns out that isn't saying much, by which I intend no offense: Keeper Trout would probably be the first to agree. *No one* knows much of anything about the taxonomy or botany of San Pedro, a common name that might or might not refer to four entirely different species of columnar cacti native to the Andes: *Trichocereus pachanoi* (which is generally accepted as San Pedro) as well as, possibly, and more controversially, *T. bridgesii*, *T. macrogonus*, and *T. peruvianus*, aka the Peruvian Torch. And then there are the countless crosses of these species, hybrids that further muddy the taxonomic waters.

Keeper Trout is the author of *Trout's Notes on San Pedro & Related Trichocereus Species*, a suitably modest title for a book whose introduction offers this warning: "We recognize the work in your hands has no authoritative merit." And this:

> We would also suggest that should our readers encounter anyone who considers themselves an expert on this ge-

> nus, or anyone who insists they know what differentiates, say, a short-spined *peruvianus* from a long-spined *pachanoi*, their best course of action is probably to nod one's head, indicating a lack of desire to argue, & leave them to their beliefs.

After spending a frustrating hour or two with Trout's book, paging through hundreds of black-and-white photos of very similar columnar cacti found in places as diverse as the Bolivian highlands, gardens in Berkeley, and the nursery department of a Target, I had the opportunity to "meet" Keeper Trout via Zoom. A slender, slightly scraggly-looking man in his sixties, Keeper spoke to me from a rustic cabin in the woods outside Mendocino. He could not have been more generous with his knowledge and enthusiasm for the whole *Trichocereus* genus. But though I've gone down some deep, dark Linnaean rabbit holes with botanists in the past, I have never ended an interview quite as confused as I was when Keeper Trout logged off my screen. My notes are an anarchy of disputatious taxonomy I see no need to inflict on the reader. But there were a few intelligible nuggets that shed some light, faint though it may be, on the mysteries of San Pedro.

The most intriguing fact Keeper Trout shared is that sometime after scientists determined that several species of *Trichocereus* contained appreciable amounts of mescaline, a notorious and wealthy cactus collector known only as DZ sought to buy up every known specimen of the plant in North America. Why?

"To prevent other people from having them," Trout said. The drug war was raging, and psychoactive plants such as peyote were among its targets. Trout believes that DZ wanted to prevent San

Pedro from being "scheduled"—added to the official list of plants it is illegal to possess and cultivate. He figured that if America's youth ever learned how easy it is to grow San Pedro and extract mescaline from it, the government would crack down on the cacti and collectors would lose their access to *Trichocereus*.

"When I first got into this in the late seventies and early eighties," Trout recalled, "it was almost impossible to find *peruvianus* or *macrogonus*"—because DZ had cornered the market. Did the strategy work? Well, to this day San Pedro has not been scheduled; anyone can grow this mescaline-producing plant without breaking the law.

Eventually, DZ lost interest in cacti; Trout heard he had moved on to collecting cowboy hats. DZ dumped his collection, flooding the market, and eventually the American landscape, with all manner of *Trichocereus*. In the years since, a perfect storm of inaccurate labeling, shoddy taxonomy on the part of so-called experts (don't get Trout started), and rampant hybridization have contributed to the confusion now surrounding what is and is not "San Pedro." Yet that confusion is not without its benefits: if the government wanted to stamp out San Pedro, it would first have to specify the names of the species to be criminalized (as it had done with *Papaver somniferum*). As a collector, however, I had hoped to pin down what species I had in my garden.

"Don't take the names seriously," Trout told me, sensing my mounting frustration. "The plants don't care what we call them."

After our Zoom session, I emailed Trout a snapshot of my cactus. He wasn't especially impressed. "It looks like the hybrid you find all over the Bay Area, probably a cross of *pachanoi* and *peruvianus*. That strain is far weaker than what shamans in Peru use, but it is what most people in the USA have known and successfully worked with." He also had his doubts about its pedigree; Shulgin, whom Trout knew, had a serious collection and probably wouldn't have bothered planting such a common hybrid.

That night Trout emailed me a recipe for preparing San Pedro. It called for a chunk of San Pedro the length and girth of one's forearm for each person planning to drink. Since only one of my candles had attained those dimensions, I decided to hold off on cooking my cactus until it had developed two hefty-enough forearms.

At this moment—that is, the moment before I harvested my San Pedro and began to cook it—my garden and I were completely in the legal clear. The act of slicing off a forearm would probably not by itself cross the line: the gardener might be taking a cutting to propagate a new cactus. But the act of cooking the cactus would change everything: as soon as I chunked up the flesh beneath the emerald skin and simmered it in water, I would be guilty of the federal crime of manufacturing a Schedule I substance. Until then, however, there was nothing to worry about.

There's something agreeable about the fact that I can make a psychedelic here in my garden without exchanging money or worrying about a visit from the police. And while extracting mescaline from that plant is technically illegal, the procedure is remarkably simple and straightforward, involving nothing more than the simmering, reduction, and filtering of a kind of cactus stock. From start to finish, the process can be accomplished without buying a thing (assuming someone gives you a cactus cutting) or having any contact whatsoever with the black market—or, now, even having to put on a mask. San Pedro: the perfect psychedelic for people in lockdown, stay-at-homes, survivalists, and skinflints.

Yet during this period my garden was not entirely innocent of scheduled plants. That's because, purely in the interest of research, I also acquired a specimen of peyote. Until recently, this diminutive cactus grew, more slowly and seemingly less happily, in a pot right next to my soaring San Pedro.

This plant, too, was a gift, from a woman I met a couple of weeks before lockdown, while visiting a commune a few miles south of Mendocino called Salmon Creek Farm. The commune had, like so many others in Northern California, fallen apart decades ago, but an artist friend of ours had recently bought the place and restored it, and Judith and I were visiting for the weekend—one of the last weekends, as it turned out, anybody went anywhere or met strangers without worrying about the virus.

A handful of the original communards still lived in the area, and on Saturday afternoon they joined us for lunch in the garden, in something of an impromptu reunion. I met a woman I'll call Aurora who had raised two kids on the commune, or had tried to—she decided it wasn't a safe place for children, and moved to a house nearby. Aurora was a gardener and a bread baker, giving us much to talk about, and within minutes of meeting her I had offers of a jar of her circa-1970s sourdough starter and, incredibly, a baby peyote plant.

Peyote had once played an important part in the life of the commune. By 1970 the Haight-Ashbury scene had curdled and the counterculture took a sharp rural turn; in Northern California especially, the commune movement was thriving. A keen interest in Native

Americans and their culture blossomed around the same time, especially among the back-to-the-landers. Here were people who actually knew how to live off the land, who were in possession of the kind of knowledge of, and respect for, nature that white kids awkwardly learning their way could only envy and try to emulate. Meanwhile the larger culture was having a reckoning with the legacy of its shameful mistreatment of American Indians, much as it is having a reckoning around racism today. Dee Brown's landmark book, *Bury My Heart at Wounded Knee*, published in 1970, told a conscience-shocking story of dispossession, cultural annihilation, land theft, ripped-up treaties, massacres, and an endless string of lies and promises broken by white America. (As Hampton Sides pointed out in his foreword to a recent edition, the book appeared at the height of the Vietnam War, not long after the revelations of the massacre at My Lai. "Here was a book filled with a hundred My Lais.")*

The counterculture embraced Native Americans†—or at least its idea of them. Indians had much to teach the communards, not only about the natural world but about living together in small tribes, and about reorienting their spirituality around the natural world. So

---

*In his *Newsweek* review, Geoffrey Wolff said that no book he had ever read "has saddened me and shamed me as this book has. Because the experience of reading it has made me realize for once and all that we really don't know who we are, or where we came from, or what we have done, or why."

†The use of the term "Native American" became widespread during this period, thought to be more respectful than "Indian," a post-colonial term based on Columbus's epically faulty sense of direction. But "Native American" has its own origins problem, since the name "American" is also a European construct, and a ridiculous one at that—based on Amerigo Vespucci's bogus claim to have discovered the continent. Ralph Waldo Emerson called Vespucci a "thief" and a "pickle-dealer at Seville" who "managed in this lying world to supplant Columbus, and baptize half the earth with his dishonest name." According to the Census Bureau, in recent years more Indigenous respondents identify as "Indian" than as "Native American." I use both terms here, depending on context, but acknowledge there is no satisfactory solution. (The Canadians have finessed these problems with the terms "First Nations" and "First Peoples.")

it should probably come as no surprise that a number of communes borrowed Native American religious ceremonies involving peyote. The communards were already familiar with the power of psychedelics, LSD especially. But LSD was a synthetic chemical—like DDT, Agent Orange, and tear gas. By contrast, peyote represented a more organic, authentic, ancient, and New World alternative, and one with an Indigenous pedigree. And at the time, it was still possible to obtain peyote buttons that intrepid hippies gathered in the Texas desert.

In 1975, in a tepee erected at Table Mountain, a neighboring commune, Aurora took part in her first peyote ceremony. The ceremony was supposedly based on the strict rules of the Native American Church. ("None of us knew about 'cultural appropriation' at the time," Aurora reminded me, slightly embarrassed at the thought.) Soon after, Salmon Creek Farm began holding its own peyote ceremonies, typically on the solstice and the equinox.

"The main attraction for us was that we felt we were here to honor the land we were living on and be in harmony with nature, and that was what we thought the Native American ceremony was about."

But then in 1982 or '83, the communards invited some actual Native Americans, up from New Mexico, to participate in their ceremony. "We were so excited! The Native Americans erected the tepee, gathered the firewood, and they had us follow *all* the rules. And we immediately could see that their ceremony was completely different from what we had been doing.

"*Oh, shit, I get it,*" Aurora remembers thinking. "What we were doing was *not* okay. We had taken their ritual and turned it into something else." (At least it didn't involve show tunes.) "But this belongs to them. We're never doing this again." The commune continued to hold peyote ceremonies at the solstice and equinox, but gave up on trying to make them "authentic."

In those days, the communards mostly used dried peyote buttons imported from Texas, but at some point Aurora began growing the cactus herself. She soon learned just how poky a plant peyote is, that it can take fifteen years to grow from seed into a harvestable button. She took me to see her collection, which she kept in a small greenhouse. The peyote cactus hugs the ground like a stone, a roundish blue-green pillow (it reminded me of a pincushion) segmented into lobes arranged in a geometric pattern, each with a little furry white nipple where the spine should be; the flower bud emerges from the center. They're modest, thornless plants, easy to overlook, yet their intricate patterning suggests a mystical object of some power.

Mature peyote plants occasionally make babies—tinier versions of themselves spun off from their edges. Using a trowel, Aurora carefully separated one of these clones from its mother, taking care to keep it attached to its taproot, which resembled a short fat brown carrot. She put the button in a small plastic pot with some potting soil and gave it to me. I brought it home to Berkeley, where, at least in the eyes of the law, it instantly transformed my garden into an "illicit drug lab."

I had a lot of questions about my new peyote cactus—horticultural, botanical, and legal—so I got in touch with Martin Terry, the botanist who had offered to give me a tour of the Texas peyote gardens before the stay-at-home order went into effect. Terry studied at Harvard under Richard Evans Schultes, the legendary ethnobotanist who specialized in the use of psychoactive plants by Indigenous cultures.

Shortly before our interview, my new cactus suffered an injury. Some animal had taken a bite out of one of its five little lobes, leaving a nasty divot in the plant and, right next to it, the missing piece of cactus flesh, evidently discarded. I was fairly sure of the culprit: a scrub jay that had nested in my hedge. I had already caught this bird in the act of systematically yanking pea shoots out of the ground in order to get at their seeds.

I spoke with Terry by Zoom at his home in Alpine, Texas, where he taught for many years in the Biology Department at Sul Ross State University. I told him what had happened to my cactus. He guessed that the bird had taken a bite of the cactus and spit it out, because the taste of the mescaline alkaloid is extremely bitter.

"It appears to have a repulsive taste to some species of herbivores," Terry said. For example, javelinas, the small pig-like mammal native to the border region where peyote grows, exhibit an aversion to its taste. Terry proved this to his own satisfaction by placing the crown of a peyote cactus on a flat rock in a place where footprints indicated heavy javelina traffic. The following morning, he found that "the peyote crown had been picked up, chewed on very slightly at the edge, and spat out again a few inches away. I believe that result to constitute one data point suggesting javelinas do not like the taste of mescaline, which puts mescaline in the category of a chemical defense." Humans, too, find the taste of peyote repellent, though they can learn to tolerate it.

These days Terry is retired from teaching but keeps busy with his work for a new organization called the Indigenous Peyote Conservation Initiative (IPCI), where he serves as the staff botanist. IPCI is dedicated to ensuring that the Native American Church continues to have access to peyote by protecting the lands where the cactus grows and, eventually, eliminating the shortage of wild peyote by cultivating it. Though the IPCI was initially funded by a California philanthropist and clinical psychologist named T. Cody Swift, a white man, the organization builds on the work of the Native American Rights Fund and the National Council of Native American Churches, members of which serve on its board and shape its agenda. Recently IPCI bought a tract of 605 acres of peyote land outside Laredo, making it possible for American Indians to pilgrimage to the peyote gardens and harvest the cactus themselves, instead of relying on the "*peyoteros*" licensed by the state of Texas to gather the cactus and sell it to them.

The licensed *peyoteros*, who are not American Indians, work quickly when harvesting the cactus, often yanking it from the ground, root and all, as if pulling carrots. Poachers do the same thing. If harvesters would instead slice off only the green button, leaving the underground stem and root intact, the plant would eventually regenerate, producing new buttons. But that takes some skill and time. Terry says that many *peyoteros* hire high school kids to do the work on a piece basis, and they can't be bothered to do it right. Nor can poachers working quickly in the dead of night.

But the shortage is the result of increased demand as well as unsustainable harvesting practices. The Church has grown rapidly in recent years, and although the precise number of members is difficult to pin down, it could be as high as 500,000. The number of peyote ceremonies is also on the rise. Unlike most religions, Native American Church services, called meetings, don't happen on a fixed schedule, but rather whenever the local "roadman," or leader, determines there is a reason to meet, and those reasons are many: to heal someone who is sick; to treat someone struggling with alcoholism or another addiction; to help a couple whose marriage is on the rocks; to send a soldier off to war; to resolve a dispute in the community; to mark a graduation or some other rite of passage.

Some think the Church needs to put limits on consumption; others, that non-Native people should be prohibited from using peyote, as they are by law if not custom. "I would prefer to work on increasing supply rather than decreasing consumption," Terry told me. He believes the only realistic solution to the peyote shortage is for the IPCI to begin cultivating the cactus: starting it from seed in the greenhouse and then transplanting it in the wild. In his view this is the best way to ensure there will be enough peyote for everyone who wants it.

There are two obstacles to this strategy. The first is Texas state law, which, though it allows for the harvesting and sale of the cactus to members of the Church by licensed *peyoteros*, explicitly prohibits the cultivation of peyote for any purpose. Terry and his colleagues at the IPCI hope to get around that hurdle by obtaining a DEA license to cultivate peyote, which is expected to happen soon. The second obstacle, which may be more difficult to surmount, is Native American belief: the peyote found growing wild is a gift of the Peyote Spirit, which it embodies; cultivated peyote is something less than that. To grow it also implies you lack faith in the Creator to provide it.

As an ethnobotanist, Terry cares not only about plants but about the ways humans engage with them, so he is sensitive to the power of such beliefs. He thinks Native American objections to cultivated peyote traces to the origin myth of its discovery.

"A woman ventures out into the desert and gets lost," he began. In some versions, she gets sick and is left behind by her hunting party. "She's in serious trouble because she's run out of food and water. Eventually she gives up, and lies down under a bush," to sleep and, possibly, to die.

"When she wakes up, the first thing she sees is a little peyote plant. 'Eat me,' the plant says. She eats, is revived, and immediately understands what peyote is about, how it works to nourish and to heal. She brings it back to her people." The predicament of the woman, abandoned and on the verge of death, is that of all Native Americans, many of whom believe, with some reason, that this cactus has saved them, whether as individuals or as a culture—but the plant as a gift from nature, rather than the chemical it contains. It probably goes without saying that San Pedro and synthetic mescaline are nonstarters for members of the Native American Church.

Terry and others at the IPCI think the ideological barrier to cultivation can be finessed. Getting the language right is important, he's found. For instance, members of the Native American Church object to the notion of a "greenhouse"—a manmade indoor structure—but not necessarily to a "nursery," a place where babies are taken care of before they're ready to go out into the world on their own. "I'm hopeful we can find a way to do this that allows the peyote to retain its cultural significance as a sacred plant."

4. The Birth of a New Religion

Peyote has been used by Indigenous peoples of North America for at least six thousand years (and possibly much longer), but its use by American Indians goes back only a century or two. The Native American Church wasn't officially established until 1918, and the religious use of peyote by American Indians wasn't documented until the 1880s—suggesting that the modern peyote ceremony is a revival of an ancient practice that had been lost, or suppressed.

Evidence for peyote's great antiquity comes from an archaeological site in southwestern Texas. Here in Shumla Cave No. 5, part of a prehistoric settlement overlooking the Rio Grande not far from where it meets the Pecos, archaeologists found three flat peyote button effigies that mass spectrometry determined to contain mescaline. Radiocarbon dating estimated the effigies had been made nearly six thousand years ago, during the Middle Archaic period. A cluster of spines from a San Pedro cactus (*T. peruvianus*) was found among artifacts in a cave in Peru and determined to be even older, by a few hundred years. These findings suggest that mescaline is the most ancient psychedelic in use. As to how it was used, or for what purpose, little is known. But New World artifacts from subsequent eras and civilizations (including the Chavin and Aztec as well as the Huichol, Tarahumara, and Zacateco) suggest that both San Pedro and peyote were revered as plants with extraordinary powers.

Zip ahead to the Spanish conquest and we find the first written accounts of the ceremonial use of both plants, much to the consternation of the colonial authorities. "This is the plant with which the devil deceived the Indians of Peru in their paganism," wrote the Spanish priest Bernabé Cobo, referring to San Pedro. "Transported by this drink, the Indians dreamed a thousand absurdities and believed them as if they were true."

The sacramental use of these cacti posed a stiff challenge to the Christian missionary's work. Centuries later, the great Comanche chief Quanah Parker—who would become something of a missionary for the Native American Church in its early years—neatly captured the Church's dilemma: "The white man goes into his church house and talks *about* Jesus, but the Indian goes into his tepee and talks *to* Jesus." How could the bread and wine of the eucharist possibly compete with a plant sacrament that allowed the worshipper to make direct contact with the divine?

By sheer dint of ecclesiastical power, was the brutal answer. In 1620 the Mexican Inquisition declared peyote a "heretical perversity . . . opposed to the purity and integrity of our Holy Catholic faith," making it the first drug ever to be outlawed in the Americas—thereby launching the first battle in the war against certain plants that continues to this day. The gravity with which the authorities treated peyote is plain from its inclusion on the list of questions priests put to penitent Indians to judge the state of their souls:

> Art thou a sooth-sayer? . . .
> Dost thou suck the blood of others?
> Dost thou wander about at night, calling upon demons to help thee?
> Hast thou drunk peyote, or given it to others to drink, in order to discover secrets . . . ?

Between 1620 and 1779, the Inquisition brought ninety cases against users of peyote in forty-five locations in the New World. The records of those cases suggest that *raíz diabólica*, the "diabolic root," was used in one of two ways. In the first, a *curandero*, or shaman, would use peyote for the purpose of healing or divination. According to Mike Jay, the author of *Mescaline: A Global History of the First Psychedelic*, "the clairvoyant power of the peyote trance was used to reveal the location of a missing object, the cause of an illness, the source of a bewitching, prognostication of weather or the outcome of battles." Peyote brought knowledge that could help solve problems. The second use was collective and ceremonial: missionaries reported scenes in which whole villages would sing and dance all night long under the influence of peyote. "To the hostile eyes of priests and missionaries these 'feasts' were no more than drunken orgies," Jay writes. "More sympathetic witnesses would reveal them as ritual practices of astonishing complexity, woven deep into the fabric of the participants' lives."

Perhaps the longest-known continual use of peyote by an Indigenous people is by the Huichol, or Wixáritari, people, who have lived deep in the Sierra Madre of Mexico for thousands of years. The ruggedness of their landscape and their isolation have protected the Huichol (and their peyote ceremonies) not only from the Inquisition but from most attempts at assimilation. But the retreat to the mountains separated them from their traditional peyote lands. So, as they have done for centuries, the Huichols make a ritual pilgrimage to a sacred site in Wirikuta, to gather peyote for their ceremonies—enough to last till the next pilgrimage.

Their ceremony, which some anthropologists believe has changed little since the time of Cortés, is much more Dionysian in character than the formal peyote ceremony North American Indians would develop in the nineteenth century. The Huichols consume sufficient quantities of the cactus to have visions. During the course of the night they will dance and sing around the fire as well as pray and laugh and weep; compared to a Native American Church meeting, it is an ecstatic affair. At dawn the ritual concludes with an animal sacrifice and a feast—blood is believed to nourish the peyote cactus.

This last practice turns out to have some basis in fact: Keeper Trout told me that a good way to boost the mescaline content of peyote or San Pedro is to fertilize the plants with bloodmeal.

The first white man to witness a Native American peyote ceremony was James Mooney, an ethnologist working for the Smithsonian Institution in southwestern Oklahoma in 1890–91. Mooney, who as a child had memorized the names of hundreds of Native tribes, dedicated his career to documenting and preserving Native American cultures before they completely disappeared from the earth—that erasure being the explicit goal of the government for which he worked. At the time, any Native religious practices deemed contrary to Christianity were outlawed in the United States. (Some of these prohibitions on American Indian ceremonies stood until the Carter administration.) Indian boys were being forcibly removed from their families, given haircuts, and sent off to government boarding schools. The avowed purpose of these institutions, in the words of the founder of one of them, the Carlisle Indian School, was to "kill the Indian and save the man."

Mooney learned to speak Kiowa and won the trust of several of the tribes that had recently been relocated to the Indian Territory that would become the state of Oklahoma. This forced move onto reservations was devastating, and disorienting, to people, many of whom had lived itinerant lives, moving with the seasons and the bison. Suddenly they found themselves dependent on government rations of beef and corn. Some Plains Indians, hunters rather than agriculturists, didn't recognize corn as human food, so they fed it to their horses.

Mooney was particularly interested in documenting Indian religious practices, old and new, and in the course of his years in Oklahoma, he learned of two new religious movements: the Ghost Dance and the peyote religion. Both these movements were pan-tribal and both were spreading rapidly across Indian Territory, but each represented a completely different response to the existential crisis facing Indian culture as a bloody and calamitous nineteenth century drew to a close.

Of the two, it is the peyote religion that has survived and flourished, but its success can't be understood without knowing something about the Ghost Dance, short-lived as it was. Mooney was one of a small handful of white people ever to witness the Ghost Dance and his account is the best we have, at least from a Western perspective. The ritual was inspired by the mystical experience of a Paiute man named Jack Wilson, aka "Wovoka." During a solar eclipse on New Year's Day in 1889, Wovoka had a vision in which God told him he had prepared a new world for the Indians, one from which the white man had been erased. Wovoka was shown a new dance that would help usher in this promised world—a return to a Golden Age before the calamity of the Europeans' arrival.

Wovoka's ecstatic ritual spread swiftly from tribe to tribe, with massive gatherings of Indians donning extravagant costumes and dancing in a vast circle while singing the new "Messiah songs." This would go on for twenty-four hours, with the participants falling into trance, "some in a maniac frenzy," Mooney wrote, "some in spasms, & others stretched out on the ground stiff and unconscious . . . while the dance goes on." Mooney likened the Ghost Dance to a revival meeting, with participants speaking in tongues and falling into a trance state, but few whites could appreciate the resemblance.

The strange new religion suddenly rolling through the Indian Territories terrified the authorities; to them, the Ghost Dance looked less like a revival meeting than a prelude to insurrection. In a panicky effort to suppress the "messiah craze," the Indian police shot and killed Sitting Bull, the Lakota spiritual leader, in December 1890, and then, after attempting to disarm several hundred Lakota whom they had lured to Wounded Knee Creek, the 7th Cavalry Regiment surrounded them and opened fire, killing more than 250 men, women, and children in one of the bloodiest massacres in American history. The Ghost Dance was no more.

A few years earlier, and in response to the same sustained campaign to uproot and destroy Indian culture, a second pan-tribal religion sprang up in Indian Country and began to spread from one tribe to another. That spread was accelerated by the policy of forcing far-flung tribes onto reservations in Oklahoma, putting them into closer contact with one another, and fostering a greater sense of "Indian identity" in the face of its oppression. Compared to the Ghost Dance, the peyote ceremony was a sedate affair, conducted inside a tepee and featuring "a certain Christian ambience," in the words of historian Omer C. Stewart, that made it much less threatening to the authorities. The meetings "carried a high moral tone such as might characterize a mission service." And since they took place inside, peyote ceremonies could be conducted quietly and out of view of white people.

Quanah Parker played a pivotal role in the Indians' abandonment of the Ghost Dance and embrace of the new peyote religion. The offspring of a Comanche chief and a white woman who had been taken captive as a young child and raised by Indians, Quanah Parker overcame the stigma of his white blood ("Quanah" means "smelly") by proving himself a great warrior. Rather than submit to life on a reservation, Parker chose to battle the government, but after he was ultimately defeated, he deftly navigated the transition from outlaw to prosperous rancher and trusted go-between with the authorities.

Parker had his first experience with peyote in 1884; he claimed the cactus had cured him of a stomach injury sustained after being gored by a bull. A pragmatist skeptical of messianic fantasies bound to end in disappointment (or worse), Parker saw in the new peyote religion a constructive alternative to the Ghost Dance, a ritual of accommodation to the Indians' new reality rather than one promising escape. (What an irony!—that the more pragmatic and acceptable of the two rituals was the one involving a psychedelic.)

Parker became a roadman, a charismatic leader of peyote ceremonies, and, in time, the Johnny Appleseed of peyotism. He traveled all over Indian Territory, bringing his bag of peyote buttons and leading meetings for the Cheyenne, the Arapaho, the Pawnee, the Osage, and the Ponca, among other tribes. When the federal government sought to crack down on peyote in 1888, threatening to withhold rations from anyone found using it, Parker successfully defended the practice before the authorities, arguing, with some success, that the peyote religion should be regarded as a complement to Protestantism rather than a challenge. It was no accident that he would talk about seeing Jesus under the influence of peyote rather than the Great Spirit.

James Mooney shared Quanah Parker's enthusiasm for the new peyote religion, which might explain why he became the first white man invited to witness a meeting, in 1891. In a series of reports he described a rigidly plotted all-night ceremony conducted around a fire in a tepee. Officiated by a roadman, a drum chief, a fire chief, and a cedarman, the ceremony leaves nothing to chance, not even one's posture: participants must sit upright and cross-legged through the night with their eyes open, staring into the fire. A crescent-shaped altar is mounded out of earth, with a large "grandfather" peyote button placed on top. Ceremonial objects, such as the gourd rattle, water drum, and staff, are always passed to the left, as is the basket of peyote buttons, which comes around several times over the course of the night; in one of the few elements of the ritual that might be called spontaneous, participants can decide for themselves how many buttons to ingest. The roadman offers prayers. Participants take turns singing songs, each one four times; the rhythm of drumming is rapid and unceasing.

At midnight there is a break, allowing participants to stretch their legs. (Few take advantage of the opportunity, Mooney noted, since doing so is regarded as a sign of weakness.) At this point prayers are said for anyone who is sick. Mooney described a powerful moment when the door flap opened and a man entered the tepee holding an infant "child sick almost to death." The roadman prayed over the man's child, after which "he left as silently as he had entered." Also, at midnight there is a water ritual that Mooney described as a "baptismal ceremony." Water is then passed around for everyone to drink.

"Each man then calls for as many peyotes as he desires to eat, and the songs are resumed, increasing in weird power as the effect of the drug deepens." This goes on "until daylight begins to glimmer through the canvas." As the ceremony drew to a close, the roadman turned to Mooney and told him he "should go back and tell the whites that the Indians had a religion of their own which they love."

This Mooney did, devoting much of the rest of his career to defending peyotism and helping to establish the Native American Church. He argued to his superiors at the Smithsonian and anyone else who would listen that the new religion promoted religious and moral inspiration as well as sobriety, alcoholism having emerged as a scourge among Indians relocated to reservations. Mooney fervently believed that the new peyote religion offered a means to rescue Native culture and identity from imminent collapse while at the same time helping Indians adjust to the strictures of reservation life. "Rather than awaiting a transformation of the world," Mike Jay writes, "it gave its worshippers a means to transform themselves from within."

The government had no interest whatsoever in the survival of Indian identity; to the contrary, its policy was to extinguish it. The new religion might not be as threatening as the Ghost Dance, but Christian missionaries determined to stamp out peyotism, which they regarded as heathen and no different than alcohol. At the missionaries' behest, Oklahoma passed the first law banning peyote in 1899, though within a decade it had been repealed, largely as a result of Quanah Parker's lobbying efforts.

Soon after, however, peyote got tangled in the politics of Prohibition; William "Pussyfoot" Johnson, a notorious Prohibitionist who called peyote "dry whiskey," took it upon himself to raid peyote meetings in Indian Country. Around the same time, another opponent of peyote, Superintendent Charles Shell of the Cheyenne and Arapaho Agency, decided he should find out for himself what peyote did to the mind. He ate some at home in the company of a doctor and was astonished to find himself having thoughts "along the line of honor, integrity, and brotherly love.

"I seemed incapable of having base thoughts. . . . I do not believe that any person under the influence of this drug could possibly be induced to commit a crime."

But Shell's unexpectedly favorable trip report did little to discourage the Prohibitionists, who along with the Bureau of Indian Affairs (operating under the sway of the missionaries) pressed for a federal law banning the cactus. Only the organized efforts of American Indians themselves, as well as the congressional testimony of white advocates such as James Mooney (and, later, Richard Evans Schultes), turned back repeated attempts to crush peyotism.

Hoping to gain the protection of the First Amendment, representatives of several tribes came together in El Reno, Oklahoma, in August 1918 to sign the articles of incorporation of the Native American Church—marking the first time that Indians officially referred to themselves as Native Americans. James Mooney played a critical role in the negotiations leading up to this event. The charter, which made explicit reference to the "peyote sacrament," stated that the Church had been incorporated "to foster and promote the religious belief of the several tribes of Indians in the State of Oklahoma, in the Christian religion."

But the battle was far from over. Legal and political skirmishes about the legitimacy of the peyote religion would continue for the rest of the twentieth century, as peyotism, having barely survived Prohibition, now got caught up in the drug war. Beginning in the 1960s, peyote meetings were frequently raided and Indians found in possession of peyote were arrested. Civil liberties organizations like the ACLU took up the American Indians' cause, and a body of law gradually developed affirming the Native American Church's First Amendment right to the free exercise of religion.

It was in pursuit of precisely this freedom, of course, that the American colonialists originally fled Europe, coming to the Indian lands they rechristened New England. That their descendants would now seek to suppress the Indians' own religious freedom was an irony apparently lost on most Americans, including the justices of the U.S. Supreme Court. In a shocking 1990 decision written by Justice Antonin Scalia, the Native American Church lost its right to practice its religion. Up to that point, the courts had held that the government could not deny one's First Amendment right unless it could demonstrate a "compelling state interest." But in *Employment Division, Department of Human Resources of Oregon v. Smith* (Alfred Leo Smith was a member of the Klamath Nation who was fired from his job when he refused to stop attending Native American Church meetings), Scalia threw out the compelling state interest standard. Calling America's religious pluralism "a luxury," he held that the criminal law and the police power must take precedence over the free exercise of religion. (As the attorneys for the Church commented, the decision, in effect, "rewrote the First Amendment to read, 'Congress shall make no laws except criminal laws that prohibit the free exercise of religion.'") The government's interest in prosecuting its war on drugs had won out over the First Amendment's protection of religious liberty.

Scalia's ruling sparked outrage in the larger religious community, which the very next day came together to ask the Court to reconsider its decision. In his opinion, Scalia had advised the Church to turn to the legislature to win back the right the Court had taken away, and within only a few years of Scalia's decision the Church did precisely that. In 1993 Congress passed the Religious Freedom Restoration Act, which restored the compelling state interest standard. This represented progress, yet it didn't guarantee a government wouldn't find some compelling interest to ban the use of peyote, especially during the drug war. Led by the Winnebago tribal leader Reuben A. Snake Jr., the Native American Church assembled a coalition and launched a campaign to press Congress to specifically protect the Church's freedom to use its peyote sacrament. On October 6, 1994, President Clinton signed the American Indian Religious Freedom Act Amendments. Henceforth, "the use, possession, or transportation of peyote by an Indian for bona fide traditional ceremonial purposes in connection with the practice of a traditional Indian religion is lawful, and shall not be prohibited by the United States, or any State." A century after the new peyote religion had sprung up on the Great Plains, the Native American Church had secured the legal right to use its sacrament.

5. *Peeking Inside the Tepee*

It's not easy for an outsider to learn exactly what the peyote ceremony means to Native Americans today, or what it has given them. Clearly a great many of them regard it as precious, even indispensable. Members of the Native American Church I spoke to credit peyotism with revitalizing and sustaining traditional Indian culture; promoting sobriety; healing diseases of both the body and mind; and creating bonds among Indian tribes who have often found themselves at odds.

But how, exactly? How does this ceremony and its psychoactive sacrament effect all this . . . personal and collective transformation? I had hoped to find out for myself by attending a Native American Church meeting in Texas in November, but that, alas, was not to be. That left Zoom. After interviewing a number of roadmen, Church officials, and members from several different tribes, I have a better sense of what takes place in the tepee, but I'm still not entirely sure I understand it. Part of that uncertainty owes to the epistemological gulf between Indigenous and Western ways of thinking about plants and medicine and "drugs." But I also encountered a deep reluctance on the part of many Natives to share—at least with this white person—exactly what goes on behind the tepee canvas.*

The reticence to discuss spiritual matters with a white writer from Berkeley should not have surprised me. Steven Benally, a Navajo roadman in his seventies who currently serves as president of the Azeé Bee Nahaghá of Diné Nation (formerly known as the Native American Church of Navajo Land), regarded me with open distrust when I asked him what I thought was a straightforward question: What had peyotism done for his people? I had reached him at his home in Sweetwater, Arizona, on the reservation, which had been hit particularly hard by the pandemic; when we spoke in May, eight people of his acquaintance had already died. Benally's affect was calm, dignified, and deliberate, but at times he flashed a fierceness that caught me off guard.

"I'm guessing you're white, yes?" Benally began. "All this information you want, what's in it for me? It's a dilemma I have, talking to you. If I divulge too much information about how peyote is good for

---

*One of the best Native American accounts of a peyote ceremony, described in detail by Leonard Crow Dog, appears in Lame Deer's *Seeker of Visions* (New York: Washington Square Press, 1979), 207–9.

this particular thing, about how it works, and give some testimonial of how this peyote heals, you might write something that creates curiosity about it among these psychedelic people." He knew I had written a book about psychedelic science, two words he had no use for.

"I'm very aware of our history, and what colonization has done to us, and the doctrine of 'discovery.'" The implication was clear: much had been taken from Indigenous peoples under the banner of "discovery" and, from his perspective, I was another in a long line of white discoverers from whom nothing good could come.

"We have been given this plant for our own needs. We must protect it for the sake of our children and grandchildren, for a future time when they're going to need it to help them survive. [Benally is a founding member of the Indigenous Peyote Conservation Initiative.] To show and tell the world how it works and what it is good for is something I'm kind of scared to do. Do you see what I mean? If there's money to be made from peyote, nothing will stand in the way." Native Americans of Benally's generation remember the 1970s fad for peyote inspired by Carlos Castaneda, which drew an untold number of hippies to the peyote gardens of Texas to harvest a sacrament they regarded as a psychedelic drug, putting pressure on the only wild population of peyote in America. Another concern is that the scientists now researching psychedelics as a treatment for mental illness will turn their attention to peyote as the source of a new drug.

"We are taught to be really protective of our medicine."

After a brief surge of indignance, I realized I couldn't blame him for being so protective of his knowledge and distrustful of me. What *is* in it for him and for Native Americans to share their ceremony, and this plant, with those who have taken so much from them?

Still, I persisted, albeit more delicately, and after a negotiation about what would remain off the record (including some testimonials to miraculous healings credited to peyote), we talked for at least an hour, about everything *except* what takes place in the tepee.

Benally believes the legal status of peyote—with Native American Church members having the right to use the plant while it remains a crime for everyone else—is exactly as it should be: "The law helps us in protecting this little peyote plant."

But if the plant is such a powerful medicine, why *shouldn't* others equally in need have the ability to use it, too?

"The great spirit gave us this plant a long time ago. Before the melting pot, other people probably had the kind of connection with nature, with a place and its plants, that we still have. They once had their own healing plants, but they've been lost.

"There are a lot of people today who are searching. They've lost their connection to the land and to spirituality. They're not satisfied with Western medicine and science and are looking for that missing link. Now, they are trying to think Indian, or think Indigenous. I understand that. But we don't want our grandchildren to end up like these people! If we don't conserve peyote, that's how they are going to end up, and then they will have to look to other peoples to find their [healing] plant. That is why you do all that you can to hang on to what you have so your kids don't end up as roamers floating out there."

Benally never used the term "cultural appropriation," but it hung in the air between us. The background to his comments was a conflict that had recently erupted between the Native American Church and a new drug-policy reform movement called Decriminalize Nature. Almost overnight, this movement had persuaded municipal governments in several cities (including Oakland, Santa Cruz, and Ann Arbor) to order local law enforcement to treat the prosecution of crimes involving illicit plant medicines such as ayahuasca, psilocybin, and peyote as the lowest priority. Until the pandemic put everything on hold, the city councils of a half dozen other cities* were prepared to vote on Decrim Nature resolutions.

The movement had single-handedly reframed the politics of drug-policy reform, beginning with the word "drug," which it scrupulously refrains from using, along with "psychedelic," another baggage-laden term. No, these were now "plant medicines," or "entheogens"—a term for psychedelics meant to underscore their spiritual uses. (Entheogen means, roughly, "manifesting the god within.") Decrim Nature has done a brilliant job of naturalizing psychedelics; in effect, reframing them as an age-old pillar of the human relationship with the natural world, a relationship in which the government simply has no legitimate role. There are now more than one hundred local chapters of Decrim Nature around the country.

To those who believe adults should be able to use plant medicines without fear of the police, the early success of the movement seemed like unalloyed good news. But the Native American Church saw things differently. Worried that the decriminalization of peyote would fire demand, drawing fresh hordes of psychonauts to the peyote gardens, the Church requested that Decrim remove peyote from its list of approved plant medicines and images of the cactus from its website.

---

*On Election Day 2020, the voters of Washington, D.C., passed a ballot measure sponsored by Decriminalize Nature. As of early 2021, Denver, Somerville (MA), Cambridge (MA), and Washtenaw County (MI) had also decriminalized plant medicines.

This put Decrim in an exquisitely awkward spot. Its supporters are precisely the kind of people who deeply respect Indigenous cultures and regard themselves as woke on all questions of race, imperialism, and colonialism. Now they had run afoul of a group—Native Americans!—whose traditions and wisdom they not only revered but sought to emulate in their use of entheogens. Yet to exclude peyote from decriminalization, or limit access to it to one race and not another, would foul the beautiful simplicity of the movement's message that there can be no such thing as a "criminal" plant.

What to do? Hoping to mollify the Native Americans, Decrim agreed to stop talking about peyote specifically and refer instead to "mescaline-containing cacti." (Even though peyote had been specified as one of the plants to "decriminalize" in the texts of the Oakland and Santa Cruz resolutions.) It did not take down images of peyote from its website, however, and published a statement on the site that only further antagonized the Indians:

"It is therefore the position of the DN movement that the divine peyotl cactus does not belong to any one people, nation, tribe or religious institution. We consider it to be Mother Nature's Gift to all of humanity, and we are firmly committed to awakening humankind to the spiritual insights and important messages that peyotl teaches to the human custodians of this planet we all share and live on."

"Decrim is a slap in the face of Indigenous people," I was told by Dawn Davis, another member of the Native American Church. Davis is a *Newe* Shoshone-Bannock and lives on the reservation in the Ross Fork Creek District in Idaho; she is finishing her Ph.D. in natural resources at the University of Idaho. The natural resource she studies is the dwindling wild population of peyote. She worries that peyote could end up on the endangered species list, which could spell disaster* for peyotism and the religion it has spawned. She brought up Decrim during our Zoom call before I'd had a chance to ask her about it.

"Now a person in Oakland has more rights to peyote than I do as a tribal member living on the reservation!" She was referring to the fact that, unlike the citizens of Oakland, Native Americans didn't gain the right to cultivate peyote under the American Indian Religious Freedom Act Amendments of 1994; they also must prove their membership in a tribe and the Church in order to use peyote.

"Gaining access to peyote was not an overnight battle, not as simple as going to a city council for a vote. That was four years of hard work, after a century of struggle to secure our right to this plant."

Davis was at her desk at home when we spoke, her young daughter occasionally darting into frame, angling for her attention. She has a round, open face framed by long black hair parted in the middle. Davis was no more forthcoming about the ceremony than was Steven Benally, but for slightly different reasons.

"There aren't a whole lot of us interested in talking about our experiences." She did tell me her parents had brought her to meetings when she was a young child, and they'd begun feeding her small amounts of peyote from the time she was twelve, a common practice. (Dawn was exposed to peyote in utero, when her mother attended her grandmother's wake while pregnant.)

"People ask what I feel during a NAC ceremony—but to me, these are the most private and intimate of experiences, and even I

*Not everyone agrees. Some think the designation would help conserve the cacti and that Native Americans would be exempt.

don't completely understand them. But it's up to me to interpret them. I don't want someone else's interpretation.

"It's hard to talk about how important and sacred this medicine is, especially to people who see the plant as a thing. To me, peyote is sentient. The plant is not a thing but a relative, an elder. I have witnessed the healing power of peyote and I want to respect that in every way I can."

Davis worries that between the rising demand for peyote from Native Americans and the flaws of the current system for supplying it, the time may come when there is not enough of the cactus left for the religion to survive. The problem is that the current system, in which four licensed *peyoteros* harvest peyote and then sell it to members of the Church, is unsustainable. Too often, they work hastily, sometimes damaging the plant so that it can't regenerate. But there are other threats, too: cattle that trample the thornless cacti; the recent arrival of wind farms on the peyote lands; other types of development; and poaching, which rises along with the popularity of psychedelics. Davis acknowledges that Native Americans themselves bear some responsibility for the shortage.

"Conversations are happening with tribes about reducing consumption. You have individuals who participate in ceremonies every weekend. I call them overeaters. I'm very mindful about how much I eat, because I know how far that medicine has traveled. But many Native Americans have never been to the peyote lands; they've become disconnected from their plant." This is why the Indigenous Peyote Conservation Initiative, for whom Davis has consulted, is so vital: it promises to reconnect Native Americans to the peyote lands, creating new opportunities for them to make the pilgrimage and harvest for themselves on the 605 acres the Church now owns.

I asked Davis about the potential of cultivation to ameliorate the shortage. Like most of the Native Americans I talked to about cultivation, she was skeptical that greenhouse-grown peyote would be the same as peyote grown in the wild. "We don't know how peyote creates its mescaline. In the wild it could be the rabbits, the juniper, the soil, a migratory bird, the rains—it could be all those things that make it what it is. I worry that by taking it out of its home, it's going to turn into something else.

"I've seen videos of Martin Terry's plants, and they're living in a greenhouse behind three sets of locks! I look at these poor plants and think, what are they going through?" However, Davis is not averse to the idea of starting cacti in outdoor nurseries and then transplanting them in the wild. "But maintaining the wild populations we have should be the number one priority."

In this, white people like me have a role to play, Davis believes, which is why she accepts invitations to speak at psychedelic conferences. Her message: "Leave peyote alone. This is not what they want to hear. But I don't believe this medicine is for everyone, or that it's all about love and peace. They can synthesize all the mescaline they want, but please leave the wild populations alone."*

After speaking to Davis and Benally, I realized that calling the use of peyote by non-Native people an instance of cultural appropriation isn't quite right. To appropriate an expression of culture—a practice or ritual, say—may or may not diminish it; the point can be argued either way. Yet the practice itself does not cease to exist by virtue of

---

*Davis later reached out to me to say that she no longer stands by her position on synthetic mescaline, explaining that she can't be sure that what is being called synthetic isn't in fact extracted from cactus. "There's not enough transparency about the process for me to be certain that won't happen."

having been borrowed or copied. That is not the case with peyote today. Here, the appropriation is taking place in the finite realm of material things—a plant whose numbers are crashing. This puts the eating of peyote by white people in a long line of nonmetaphorical takings from Native Americans. I was beginning to see that, for someone like me, the act of *not* ingesting peyote may be the more important one.

Not all the Native Americans I spoke to were quite as reluctant to talk about what happens in the tepee, or even as hostile to the idea of inviting a white man to observe the ceremony provided he "comes in the right spirit." Sandor Iron Rope is a fifty-one-year-old Teton Lakota from the Black Hills of South Dakota, president of the Native American Church of South Dakota, and a central figure in the IPCI. He drove into Rapid City for our call; the internet connection on the reservation would not support a Zoom session. A gentle man, Iron Rope was disarmingly open and willing to go places in our conversation that Benally and Davis would not. When I asked him if he could take me into the tepee during a peyote ceremony, he paused, gathered up his thoughts, and gave it a try, with the warning that some of his words and concepts might be lost on me. Here is some of what he said:

> If you want to go into the tepee, you would first have to change your mind-set. In the Indigenous perspective, we are here upon Mother Earth. We feel the wind and the wind talks. The sun comes up in a certain direction and it goes down in a certain direction. And so we build an altar on the ground made from Mother Earth and in the

shape of a crescent moon. And we know Grandpa Fire is going to talk to us and commune with us, so we build the fire in a prayerful manner, making offerings along the way. The four elements—the earth, the fire, the water, the air—are going to come into the ceremony at some point in time. And then there is the plant, set upon the altar.

Some people call it the flesh of our ancestors, because that's what it is, you know, and at the same time it's a spirit. Different people have different experiences with the medicine. It talks to you at different levels: about what it is you need to see, what it is that you need to feel, or experience. The medicine knows you before you even know yourself. It is like a mirror. When people get up and look in the mirror, they can fix themselves, brush their teeth and see if they look okay, you know, presentable for society. But this medicine is a mirror that allows you to see inside yourself, into the core of your heart and spirit. The peyote knows you.

So when you start to think about something, maybe something that needs healing, what you're thinking about, what you're saying, the medicine can hear you. It's not like taking out the DSM, and getting a diagnosis. It's our way of life, talking to things and realizing the life force in all things.

Often in a meeting somebody will say, Why are we gathered? We're gathered here because I need help with this problem. It could be an illness, a divorce, domestic abuse, alcoholism. I want some prayers for this reason. That person will sit in a certain spot.

The tepee represents a family, and a home. The poles that hold it up represent the woman, the foundation of the home. And then the covering represents the male protecting the female, and the fire inside. The fire is grandpop, and the flaps represent grandma, the two of them guiding the family prayer from a long time ago. And those little pegs that hold the tepee down, those are all your children. So when you go into the tepee you are going into that spiritual family for help, for prayers, because we are all related whether we want to be related or not.

People may wander off during their meditation; they will see things and hear things and smell things, but the intercessor will remind people they are there for a purpose and bring everyone back to that purpose. The songs and the prayers and the drumming help focus everyone on the purpose.

The concept of a family praying together—this is what the government suppressed and broke up, when it sent our children to boarding school, cutting off their hair, which is sacred. To lose their hair was to lose their spiritual identity. So there was a lot of healing that was needed after that, and when alcohol was introduced on our reservations. Alcohol was stealing the spirit of our people. And then came in many other things, many kinds of trauma. But it was a spiritual battle at the beginning [to defend the peyote ceremony] and it is a spiritual battle still.

One day you may sit beside us in a teepee somewhere and you'll realize a little bit about what we're talking about.

Sometimes, sometimes if you respect something, you just have to leave it alone. You know, my dad served in the war and when I was growing up, he had a firearm in his closet. And on his bedside there were these beads, made from seeds, that he used to make craft items. As a little boy I would go in there, put my finger in those beads and move them around. One day when he came home, he said, "Hey, who was in my beads?!" I didn't want to say it was me. And after he caught us a few times, whenever we'd go into his room, we knew we couldn't touch his beads, so we just looked. That's all.

Sometimes the best way to show your respect for something is to just leave it alone.

Sandor Iron Rope's words brought me as close as I had come to a peyote meeting, and they may well be as close to a tepee as I will ever get. And, as Iron Rope had predicted, there was much in his account I couldn't completely understand.

I found some illumination in an academic book: Joseph D. Calabrese's *A Different Medicine: Postcolonial Healing in the Native American Church*, published in 2013. Calabrese is a medical anthropologist and clinical psychologist who spent two years in the Navajo Nation, working as a clinician and observing as an anthropologist for his dissertation. During his time in Arizona he attended several peyote ceremonies, and his observations helped me make sense of several things Sandor Iron Rope had said. So here, for what it's worth, is one white man's take on peyotism, a look at an Indigenous practice through the prism of Western concepts of psychology and anthropology.

Calabrese found that many Navajos share Sandor Iron Rope's belief that peyote is an omniscient spirit, capable of seeing through people and somehow "knowing" them better than they know themselves; it has the power to bare one's faults and force a person to confront them. Peyote functions in the lives of Church members much like a superego; he suggests that the plant has a gaze. Children are socialized in this belief, taught that "the Peyote Spirit knows his or her activities even in the absence of parents." Conceiving of a plant as an omniscient spirit might seem fanciful, but how different is that, really, from a psychological construct like the superego—an inner voice that recalls us to the moral and ethical strictures of our society?

What I found striking in Calabrese's account is that we have in peyote a "drug" that, instead of undermining social norms, actually reinforces them. "The Native American Church arose as a revitalization movement," he points out, "focused on personal healing, rebuilding community, harmonious family relationships, connection with the Divine, and avoidance of alcohol." Compared to psychedelics in the West in the 1960s, peyote's role in the Native American community is notably conservative. (Yet another reminder of the critical role of set and setting in any psychedelic experience.) The use of peyote in the Native American Church gives us a moral model of drug use.

That such a model exists (and it exists in other traditional cultures as well) requires us to reconsider the whole concept of "drugs" and the moral failings we associate with them. In the West, our understanding of drugs is organized around ideas of hedonism, the wish for escape, and the desire to dull the senses. Early white observers of peyotism often assumed Indians used the drug as a painkiller, Calabrese writes, when in fact "it tends to increase the intensity of sensations rather than deaden them." A psychedelic experience can be hard work, the very opposite of what most people expect from illicit drugs. Westerners also tend to put medicine and religion in separate boxes, but for Native Americans (as for many traditional cultures), religion is foremost about healing. The conflation of the two has been formally recognized by the Indian Health Service, which now covers the cost of peyote meetings (and sweat lodges) for the treatment of certain illnesses. Hard to imagine, but there is a "client service code" for a religious ceremony with a psychedelic sacrament!

What peyotism chiefly heals is trauma in its various collective and individual manifestations, the enduring legacy of official policies that sought nothing less than "the destruction of Native American cultures." Calabrese reminds us of the historical moment when the new religion began to spread across North America: soon after

Indians had been forced onto reservations and the Ghost Dance had been viciously suppressed. "Instead of focusing on a transformation of the world through the disappearance of the Europeans," Calabrese writes, peyotism "focused on personal transformation that would allow one to survive in the post-conquest situation, build a stronger community, and avoid forms of postcolonial disorder like addiction to the White Man's alcohol."

How does the peyote ceremony effect these transformations? Calabrese proposes a psychological explanation that a Native American would no doubt regard as reductive, but which seems plausible to me. Like other psychedelic compounds, the mescaline in peyote induces a state of mental plasticity, one in which you are highly suggestible and therefore open to learning new patterns of thought and behavior. While in this trance state, rigid narratives about yourself ("I can't get through the day without a drink"; "I am worthless"; and so on) tend to soften until it becomes possible to construct new ones, typically narratives of transformation or rebirth. Apart from the group setting, this model closely resembles "psychedelic therapy" as it is being practiced today in the West.

But the group setting here is critical. The fact that the healing process is unfolding within a community, with everyone listening to the same music and prayers, gazing into the same fire, and experiencing the same shifts in brain chemistry, serves to reinforce the individual's new narrative, as does the fact that the attention of the group is fixed on the recipient of its prayers. It sounds a bit like a meeting of Alcoholics Anonymous, where stories of transformation and rebirth are crafted and then cemented by the approbation of the community. Except in this case the power of the ritual is immeasurably enhanced by the altered state of consciousness all share.

For me, any inquiry into the peyote ceremony would feel incomplete without landing on some such explanation, though I can appreciate why Native Americans like Dawn Davis or Sandor Iron Rope might not buy it. Early in my research, I interviewed an attorney, a white man, who had played a key role in helping the Native American Church secure its right to use peyote. Jerry Patchen has attended more peyote ceremonies than he can count. In an email he recalled one that had left him perplexed about something that had happened during the night. So, in the morning, after the ceremony had concluded and everyone was milling around the tepee, he asked a young Navajo for an explanation.

"That is the problem with you whites. You always want to know everything. We just experience it."

6. *An Interlude: On Mescaline*

It was around this time that fortune delivered to my door two fat capsules of mescaline sulfate. The gift culture is very much alive in the psychedelic community, I've found, and a friend who knew of my interest in mescaline had somehow procured a dose for me. He knew the chemist who had made it, allaying any worry that it was actually LSD or some other counterfeit, as can sometimes be the case with mescaline. Though I hadn't yet tried San Pedro or peyote, I wondered how pure mescaline would compare. I wondered if my experience would rhyme with Aldous Huxley's. I wondered all sorts of things, but no amount of advance wondering prepared me for what was in store.

The time and place I chose for my trip seemed ideal: a benign summer's day in a house built on stilts directly above a body of salt water. The bay, its moods and patterns shifting with the breezes and the tide, filled the windows of the house and lapped at the piers supporting it. I had only a single dose, so Judith agreed to sit for me. I swallowed the two capsules at 9:00 a.m. The onset of mescaline can be excruciatingly slow, so we spent the first hour walking along the shore, a pleasant enough interlude until I started to grow impatient. "Good mescaline comes on slow," Hunter Thompson wrote in *Fear and Loathing in Las Vegas*. "The first hour is all waiting, then about halfway through the second hour you start cursing the creep who burned you, because nothing is happening . . . and then ZANG!"

It was more gradual than that for me—there was no ZANG. When I first felt the mescaline come on, I was sitting outside on the deck reading while keeping an eye on two bright yellow heads slicing through the rippling water—a pair of strong swimmers. I had glanced up from my book when I suddenly felt a wave of revulsion, almost a nausea, for print. *Why would anyone ever want to read? Work to tease meaning from all these ugly black marks?* Suddenly the whole enterprise seemed absurd. No, what I wanted and needed to do now was not to read but to *look*—at the dark blue water, at the yellow heads carving lines through it, at the grain and the stains in the cedar boards cladding the house. It was incredible how much there was to see! The pelicans lumbering over the water before slowly climbing into the sky. The diamond reflections of sunlight glancing off the ripples in the bay. The crazy shade of chartreuse in Judith's socks. I was captivated by it all, and could not imagine ever wanting to do anything but devour with my eyes all that there was to see.

I tried, recalling Huxley, to invest a few minutes studying the creases in my pants, but they weren't the least bit interesting. (Maybe because I was wearing shorts?) Yet I did recognize the quality of total absorption in the material world that Huxley had described. Any desire to get up and move was gone; there was too much to examine right here. I wrote: "There is enough here. To see, to understand, to experience." And then: "A sufficiency of reality."

The word "sufficiency" appears in my notes several times that day and it holds a key, I think, to what was distinctive about the experience. To say mescaline immersed me in the present moment doesn't quite do it. No, I was a helpless captive of the present moment, my mind having completely lost its ability to go where it normally goes, which is either back in time, following threads of memory and association to past moments, or forward, into the anxious country of anticipation. I was firmly planted on the frontier of the present and, though this would soon change, there was nowhere else I wanted to be, or anything else I needed from life in order to be content. Whatever was in my field of awareness—this sumptuous feast of reality!—was sufficient.

I wondered if perhaps I had found a hidden path out of the labyrinth of anxiety in which the virus and the fires had trapped us, that simply by lowering the horizon of my attention from the future—for the virus and the fires existed mostly there for us—I had recovered some of the beauty and pleasure in living that had been lost since the pandemic. There was a spaciousness to this present that felt like the perfect antidote to the shrunken-world claustrophobia of lockdown. Was this what it meant to become a king of infinite space?

I drank in the objects of my attention like a person who had suddenly developed an unquenchable thirst for reality. I couldn't get enough of the herringbone pattern of the water as the tide turned; the dinghies and shorebirds diligently plying the bay; the fantastic multiplicity of greens forming the far shore, sandwiched between these two great slabs of blue, one sea, the other sky.

To an extent, this is what all psychedelics do—not so much change how we feel inside (as stimulants or depressants reliably do) as imbue the world around us with never-before-appreciated qualities. On psilocybin or LSD, the objects of our attention are liable to come to life and transform before our eyes: a garden plant, suddenly sentient, might return our gaze, or a chair might take on a personality and turn malevolent. Very often on psychedelics, objects become something much more than themselves. They point, often to somewhere beyond the known world, to another plane of existence. And, sometimes, we can follow them there.

But this wasn't like that. These objects did not point. No, they were emphatically themselves—and more themselves than they had ever been. I made a cryptic note—"haiku consciousness!"—but, thinking back, I have a pretty good idea what I was trying to get at: everything in the world that day acquired this Zen-like quality of bare presence, a kind of immanence.

The poet Robert Hass has written about this aspect of haiku, which he traces to the fact that in Buddhist cosmology, there is no creator and therefore no higher plane of meaning to which nature refers. (Though Native Americans speak of the "Great Creator," nature to them is also complete in itself, embodying rather than signifying spirit.) By contrast, in the Christian conception of things nature is fallen; later, with romanticism, nature can offer redemption, serve as a means of transcendence. But either way, what nature does in our culture is point. It is encumbered by the meanings we put on it.

The poet who has done the most work scraping all that meaning, symbolism, and Judeo-Christian crust off the natural world is William Carlos Williams, who I decided that afternoon is the patron saint of mescaline. (In contrast, the patron saints of LSD, ayahuasca, or psilocybin are the visionary poets: Blake, Whitman, Ginsberg.) More than once, Williams managed in his poems to evoke on the page the bare actuality of things, never more effectively than with his wheelbarrow:

> so much depends
> upon
>
> a red wheel
> barrow
>
> glazed with rain
> water
>
> beside the white
> chickens

Rereading Williams in the aftermath of my day on mescaline, poetry that had always left me a little cold, I felt a shock of recognition. *These are the eyes I was seeing with!* Here was the sheer "isness" of the given world and its objects at a particular moment in time. Haiku consciousness.

And yet at the same time there is something here—both in the poem and in the world as it appeared on mescaline—that, for all its beauty, feels almost more than a mind can bear. Is it the poignancy or the transience or what? I'm not sure. But as the mescaline intensified, my initial delight in the is-ness and immanence of objects gave way to a shiver or shadow I couldn't quite account for—until a phrase, from another poet, popped into my head: "the immensity of existing things."*

It was this—the immensity of existing things—that began to overwhelm me during the next phase of the day, as peak intensity approached and things took a darker turn. I neglected to mention that Hamlet's claim to be king of infinite space was conditional: the very next line is "were it not that I have bad dreams." Here they came. Now it felt like this was more reality than I could handle. Wide open, my senses were admitting to awareness exponentially more of everything—more color, more outline, more texture, more light. It was, to quote from Huxley, "wonderful to the point, almost, of being terrifying." Indeed. I felt as though things could easily tip over into terror.

Huxley's trip had convinced him that the function of ordinary consciousness is to protect us from reality by a process of reduction or filtration—he spoke of consciousness as a "reducing valve," and the metaphor had never seemed more apt. Throwing open the doors of perception was wonderful, in the literal sense of the word, but without the usual filters of consciousness there came the fear "of being overwhelmed, of disintegrating under a pressure of reality greater than a mind, accustomed to living most of the time in a cozy world of symbols, could possibly bear."

This is where I now found myself, and for a moment, it felt like a kind of madness. My first-person subject was still present, but it

*The line comes from "Esse," a prose poem by the Polish poet Czesław Miłosz.

lacked all volition, was too passive to defend itself from the assault of reality, of infinitude. So I closed my eyes, hoping to stanch the torrent of sensory data inundating my awareness. This provided a respite, but only briefly. Now I saw an intricate pattern of bodies entwined and dancing on a vertical scroll, reminiscent of Hindu miniatures in tantric or yoga poses. When I then tried to empty my mind by meditating, the "I" that was meditating wasn't recognizable as my own—it kept changing, one stranger after another taking turns meditating in my mind. The one I remember most clearly was a young Latin American woman in a white peasant dress who seemed to have some connection to the Indigenous mescaline users I'd been reading about and interviewing. Eventually, eyes closed proved even more overwhelming than eyes open; now, instead of the senses and outward reality, the inner floodgates of emotion opened wide, admitting cresting waves of sadness for people I had lost or fallen away from, and a boundless pathos for all the people, known to me or unknown, suffering now and before and in the times to come, more suffering than anyone could possibly hold in his head without it cracking open. It seemed possible the admission of so much suffering could kill a person.

I opened my eyes again, having decided I stood a better chance of withstanding the flood from the open valve of the senses than that of emotion, memory, and imagination. Never had my eyelids felt so crucial—powerful technologies for changing the channels of consciousness.

What was happening in my brain?! The notion that there is so much more out there (or in here) than our conscious minds allow us to perceive is consistent with the neuroscientific concept of predictive coding. According to this theory, our brain admits the minimum amount of information needed to confirm or correct its best guesses as to what is out there or, in the case of our unconscious feelings, in here. These top-down predictions of reality and prior beliefs are a bit like maps to sensory and psychological experience, and as long as they represent the actual territory well enough for us to navigate it successfully, there's no need to flood the system with lots of unnecessary detail. Natural selection has shaped human consciousness not necessarily to scrupulously represent reality but to maximize our survival, admitting only the "measly trickle"—Huxley's phrase—of information needed for us to get by rather than the full spectrum of what there is to perceive and think.

Psychedelics seem to mess with this system in one of two ways: In some cases, the brain's predictions about reality go haywire, as when you see faces in the clouds or musical notes leap to life or something happens to convince you you're being followed. Common on LSD or psilocybin, this kind of magical thinking might occur when top-down predictions generated by the brain are no longer adequately constrained, or corrected, by bottom-up information arriving from the world via the senses.

But if Huxley's account and my experience are representative, then something very different happens in the brain on mescaline. Here, the bottom-up information of the senses and the emotions inundates our awareness, sweeping away the mind's predictions, maps, beliefs, and "cozy symbols"—all the tools we have for organizing the inner and outer worlds—in what feels like a tidal wave of awe.

The overwhelming peak of the experience didn't last long, fortunately, and eventually I found my footing, allowing me to navigate all the information coming in without capsizing. Mescaline goes on and on—it is, assuming you're enjoying it, the most generous of psychedelics—and I settled in for the twelve-hour ride. Now, having regained a measure of mind control, I could choose to go deep on whatever I looked at or thought about. Later that afternoon I got chatty, and enjoyed being close to Judith. Together we listened to music, and I could hear more in the notes and their arrangement than I ever knew was there. The late-afternoon sun was raking the house, which inspired thoughts about shadows and the way they offered commentary—ironic, humorous, sarcastic—on the objects that cast them—their putative masters. What about musical notes—could they cast shadows? I listened for them. (*Definitely!*) I studied the bay out the window and registered every minute shift in color or mood. My heart felt opened up by the molecule, the windows of my senses, too: there was so very much here to savor, being in this place and moment by Judith's side.

At one point that afternoon I entertained a slightly macabre thought: How exactly would this place and moment in time feel if I was experiencing it in the knowledge my death was imminent? Weeks or even days away? All of it would feel infinitely precious and poignant. Every detail of the scene I would prize as a gift, to be tightly held in the embrace of the senses: the blush of the fragrant apricots in that blue bowl, the reflection of the clouds in the glass of the water at ebb tide, the plaintive cry of a gull reaching us from across the bay. How it would feel, I realized with a jolt, is exactly as it feels right now.

So why not feel like that always? Well, it would be exhausting, surely, to turn life into this sort of unending observance. Ordinary consciousness probably didn't evolve to foster this kind of perception, focused as it is on being—contemplation—at the expense of doing. But that, it seems to me, is the blessing of this molecule—of these remarkable cacti!—that it can somehow crack open the doors of perception and recall us to this truth, obvious but seldom registered: that this *is* exactly where we live, amid these precious gifts in the shadow of that oncoming moment.

I made a note so I wouldn't forget what I'd learned after the mescaline wore off: "Had mescaline shown me the door in the wall?" If so, then the door was—just as Sandor Iron Rope had tried to tell me!—more like a mirror, for everything I needed to learn was not on the other side of it but right here in front of me, and it had been right here all along.

7. Learning from San Pedro

The Indigenous people I had interviewed had no interest in mescaline the molecule or the sort of experience I had had on it; for them, the power was in the cactus, whether peyote or San Pedro, and specifically in the cactus as it manifested its power in ceremony.

I was more eager than ever to participate in a ceremony. Yet beyond the logistical problem of getting to Texas and spending a night in a crowded tepee during a pandemic, there was now the injunction of the Native Americans to consider: to respect the practice of peyotism, as a white person, meant leaving peyote alone. Flying to Peru was out of the question—the country had been hit particularly hard by the virus—and Don Victor's next trip to Berkeley was who knows when. But I had a lead on a "medicine carrier" who had trained with him. She now led Wachuma ceremonies (the term "San Pedro" never crosses her lips) in a place that could be reached without getting on a plane. We began to talk and then to meet, outdoors, in her garden and mine.

Taloma, as she asked me to call her, fell into medicine work in her thirties. At the time, her marriage had just fallen apart. "I was not in a good place. I was living in cheap motels, eating fast food, alone." One day, driving through Big Sur, Taloma spotted the sign for Esalen, the legendary retreat center where the human potential movement got its start. Curious, she pulled in but was turned away at the gate: only participants in workshops were being allowed on the property. She left with a copy of the catalog. During a stop at a town a few miles down the road, Taloma managed to lock herself out of her car. Waiting hours for the tow truck to arrive, the only thing she had to read was the Esalen catalog.

"It was full of all these woo-woo esoteric things," she recalls. Taloma was not exactly the Esalen type. She had never used pot, much less anything stronger, and considered herself too much the rationalist to believe in "souls or energies." Yet Esalen, with its organic food and hot baths, seemed like the perfect refuge. So Taloma signed up for a weeklong workshop—"Healing the Child Within." The experience set her on a journey of self-healing that, in time, led her to her calling: healing others with the help of what she calls "the master plant medicines."

Taloma wound up living and working (in the garden) at Esalen for several months; "this powerful, sacred, healing land" went to work on her. "It saved my life," she told me. While in Big Sur, she was introduced to the "red path": working with a Native elder named Little Bear, Taloma did a series of vision quests in the Santa Lucia Mountains behind Big Sur, fasting alone in the wilderness for four days, then seven, and eventually even longer. She participated in sweat lodges.

On the day she left Big Sur, Taloma had a near-death experience: the Jeep she was riding in flipped over three times on Route 1 before nearly plummeting into the ocean. She remembers finding herself in a tunnel with a light in the distance, before returning to consciousness. She had broken her neck and required extensive surgery to regain mobility. It was during a painful, years-long convalescence that Taloma discovered the healing power of the psychoactive plants used in Indigenous ceremonies—ayahuasca, peyote, Wachuma, tobacco. She was on "the medicine path."

With her high cheekbones and long, straight hair parted in the middle, Taloma could be mistaken for Native American. In truth, she is mixed-race, primarily Japanese American, with a trace of Native American ancestry, according to family lore. But while Taloma often mentions that fact, she also takes pains to remind people "I'm not Native American. I didn't live that struggle, and wasn't brought up in that culture." Her reverence for Indigenous culture is such that wherever she finds herself, she will seek the blessings of local Native Americans before holding ceremonies on their land.

In the years since she first embarked on the medicine path, Taloma apprenticed herself to elders in two different lineages: the Sacred Fire of Itzachilatlan, a fairly new spiritual movement based in Mexico that seeks to reunite the Indigenous cultures of North and South America by combining their ceremonies and plant medicines; and the traditional Wachuma ceremony of Peru, which she learned from Don Victor and his teacher, Don Agustín. It was only after twenty years of apprenticeship that Taloma felt ready to conduct ceremonies and offer medicine herself.

Among all the plant teachers she's worked with, Wachuma occupies a special place. "Every plant has its own spirit," she told me.

"I've connected to Wachuma because of its indomitable will to survive." It's true! Chop off a piece of Wachuma cactus, leave it anywhere—on the ground or on pavement, in the sun or darkness—and it will soon sprout a new cactus from the amputated limb. As long as it doesn't freeze hard, the plant will grow anywhere: city or country, in the mountains or at sea level, indoors or out; is happy to be watered but will go months without a drop; will send up new growth from any cut or injury and, for a cactus, grows fast—easily a foot a year. Though it flowers spectacularly and can produce seed, its principal reproductive strategy would seem to depend on disaster: getting whacked by machetes or toppled by the wind. Whatever befalls this plant it takes in stride, just another opportunity to send up new life. Compared to poky and vulnerable peyote, Wachuma *is* indomitable.

"This is the kind of medicine I want to bring to people," Taloma says. "It knows the energy of the city, the planes overhead, the sirens in the street, the wi-fi and cellphone waves we can't escape. Wachuma knows what we're dealing with. It's also a gentle and heart-opening plant. I feel strongly that it's the right medicine for this moment."

Taloma hadn't conducted any Wachuma ceremonies since the pandemic, but she had one planned for late August, and I was excited when she invited Judith and me to participate. In deference to the virus, the overnight ceremony would take place outdoors, with proper social distancing; we would wear masks, and drink the medicine from paper cups instead of a shared ceremonial chalice. And everyone would have to take a test for the coronavirus a couple of days before the event.

The week before the ceremony, Judith and I purchased mail-order COVID-19 tests. We bought new sleeping bags in case the night was a cold one. In a long Zoom chat we "met" the dozen or so people in Taloma's *allyu*, or medicine circle, and shared our intentions for the ceremony. We would drink three cups of Wachuma over the course of the night. Two weeks before the ceremony, I joined Taloma and two of her helpers as they harvested long limbs of Wachuma from a large planting she cared for, using pruning saws to slice through the unexpectedly tender flesh. We made a date to cook a few days before the big night.

And then, on the Saturday night the week before the scheduled ceremony, an immense lightning storm swept across Northern California. A spidery tangle of bolts completely filled the western sky, startling millions of people awake, all of them with the same terrifying thought: fire. In the space of an hour, more than a thousand strikes had hit the parched late-summer landscape, igniting hundreds of fires. Within days the smoke had dimmed the sun and yellowed the sky, and on Wednesday morning, Taloma sent around a long email calling off the ceremony.

"We are waking up to a new day," she wrote. "Spirit has spoken loudly with an incredible lightning storm that has set fires across the state. . . . Any who have the time, space, energy to send prayers out to all who are in fear and anxiety right now, for their physical safety, for the animals and land . . . right now . . . please do so."

And that was that. I know this is an embarrassingly small-minded way to think about natural disasters that had upended so many lives and, by now, incinerated thousands of homes and some four million acres of forest, but I couldn't help feeling that I'd been thwarted yet again. Though Taloma wrote that she hoped to reschedule soon, now that fire season was upon us, the ceremony might not be possible until the rains came, when holding it safely out of doors would be difficult if not impossible. I needed a Plan C. But what *was* Plan C?

8. Drunk at the Wheel

With the fires something changed. The accumulation of disasters was taking its toll, not only on my plans but, now, on me. Somehow, I had managed to keep my spirits up through the first six months of the pandemic. But now the invisible threat had been reinforced by a second threat you could see and feel: a fine ash was falling from the sky, dusting plant leaves and cars and entering our bodies. COVID had rendered the outdoors the safe place; now the fires were forcing us back indoors, to compulsively check websites that assessed the degree of peril it had become to breathe. Our world, already made small by the pandemic, now contracted still further.

A "red flag" warning went into effect. That meant we were to prepare a "go bag" in case of an order to evacuate, which could come at any time. So we filled a small suitcase with essential items, though what exactly qualified as truly essential changed every time we tried to decide the question.

When I embarked on this project several months ago, it was mostly curiosity that drove me. What could I learn by tracing mescaline's story, and having an experience or two with it—about the cacti, about Indigenous religion, about the possibilities of consciousness? I hadn't gone into this looking to be "healed," whatever that meant. Yet for Taloma that was the whole point of working with Wachuma. What else is medicine for?

When Taloma first asked me to formulate a prayer in preparation for our ceremony, I came up with something that was more academic than therapeutic: *What could Wachuma teach me about my mind?* Taloma didn't say so, but I could tell she was disappointed. I knew she thought (rightly) I lived too much in my head, so I revised the prayer to make it a tad more personal: I wished—okay, *prayed*—to be less in my head and more in my heart, to be more present to my emotions.

These words—indeed the whole contemporary vocabulary of healing—sit awkwardly on my tongue. But after the fires came, I lost some of the mental energies and momentum that had propelled me through the first months of the pandemic without the friction of despair I now began to feel. I started to wonder: Could Taloma possibly be right? Could this plant help us find a path through the serial catastrophes of this terrible year?

"Trauma" is a word in heavy rotation these days. Taloma talked endlessly about it, how trauma "settles in your body" and "blocks energy" and, if it's not addressed or acknowledged, can fester, leading to physical illnesses such as cancer, as "dis-ease" turns into disease. An unrecognized trauma can also lead to addictions, it's often claimed, as people seek to "self-medicate" with substances or compulsive behaviors. Healers talk about how plant medicines often "surface hidden trauma" so that they can be "worked through." *How* often? I wondered. Wasn't trauma by definition an exceptional event? Now it seemed like everyone suffered from some trauma; they just didn't know it yet.

Here in the midst of the pandemic, the fires, and the darkening political season, I began to think that my skepticism might not be supportable. I had stumbled across a psychologist quoted in the newspaper explaining that trauma is not necessarily a discrete, dramatic event. What trauma is really about, she said, is the sense of helplessness we feel when we're assailed by unpredictable forces beyond our control. Is this not our reality now? And then, in an image that I can't shake, she said: "It's like we're in an endless car ride with a drunk at the wheel. No one knows when the pain will stop." Thousands of readers must have recognized themselves in that image, white-knuckled in the back seat of that careening car. I know I did.

Just when Taloma's email canceling the ceremony popped up in my in-box, I had been trying to write a new prayer, this one asking frankly for help.

9. *Plan* C

"Wachuma doesn't heal you by itself," Taloma said. "Its power is in its subtlety. Unlike ayahuasca, which will grab hold of you and take you on a journey whether you want to go or not, this medicine doesn't put anything inside of you. But if you invite it in, it helps to reveal what is already there, and in that way engages you in healing yourself. I have seen miracles."

We were sitting around a table in a garden, observing the proper social distance, while Taloma showed me how to cut up cactus to make a small batch of Wachuma tea. After the ceremony was canceled, I had asked her if she would teach me how to cook Wachuma, and she agreed to a tutorial.

Taloma began by taking a bundle of dried sage from a purse and lit it; she then smudged the plant, the knives, and then us with fragrant smoke. There were two ways to cook cactus, and Taloma showed me both. The first, more painstaking method calls for cutting the spiky plant into foot-long lengths with a knife, and then systematically removing its defenses. First the spines, by cutting a tiny notch around each areole and then scooping them out, taking care to remove as little of the precious flesh as possible. Next you stand the piece of cactus on end and, using a long knife, carefully slice down the length of each rib, separating it from the woody white core, which is discarded.

After cutting the long triangular ribs into more manageable lengths, you remove the cuticle—the tough, semitransparent layer of skin that, along with the spines, protects the plant's watery flesh from its unforgiving environment. This was the painstaking part: gaining sufficient purchase on an edge of cuticle, either with a paring knife or a thumbnail, in order to slowly peel it off in strips. Shorn of its defenses, the cactus's flesh is surprisingly tender and moist, like a soft cucumber. It had the puckering bitterness of any plant alkaloid; think of oversteeped tea, but nastier.

As I sat across the table from Taloma on a benign summer afternoon, learning how to slice and dice cactus flesh, the work felt much as cooking in the company of others always does: pleasant, desultory, productive. The scene made me think of chefs prepping vegetables for a stock, and in a sense that's exactly what was happening. The work occupied the hands but didn't demand one's full attention, so we chatted—about the fires, other recipes, Don Victor. What the work didn't feel like was breaking the law. If I had any worries at all that afternoon, it was that I probably wasn't worried enough.

The second method Taloma showed me for cooking the cactus was both easier and more gratifying, though it works only with a fairly young plant that hasn't yet developed a woody core. After removing the spines from a foot-long length of cactus, you simply slice it through the center, as thinly as possible. This yields dozens of paper-thin six-pointed stars, their bright chartreuse coronas fading to snowy white at the center.

Taloma piled these stars in a tall spaghetti pot, filled it nearly to the brim with water, and put it on a burner. This is when the domestic cooking scene gave way to something more ceremonial. Taloma lit her sage and smudged the pot of cactus with its smoke. Then she bent over the pot, looking down at its bright green stars bobbing in the clear water, said a prayer, and, in Spanish, began to sing.

Before Taloma left, she offered these instructions: bring the pot of cactus to a rolling boil and cook it for three days, more or less, being careful to add more water whenever the level in the pot dropped below a couple of inches. When the stars turn from white to translucent, it is ready. Cool, then filter the mash through a fine cloth, put the pot back on the stove, and reduce the liquid by half. Pour the tea into Mason jars and store in the refrigerator.

When I finally "met" Don Victor, Taloma's teacher, he was in Cuzco and I was in Berkeley. Zoom wasn't working so we were on WhatsApp, reducing each of us to a postage stamp on an iPhone screen. Taloma served as interpreter, a challenging task since Victor spoke in torrents, shuttling back and forth between a world we shared (life under pandemic) and one we most definitely did not. That realm had its own intricate cosmology, based on higher and lower frequencies of vibration, other dimensions of existence, past lives, and sacred places, all of which seemed to be located somewhere in Peru. Honestly, I was lost a lot of the time, and when I wasn't, I felt as if I had stumbled into a world dreamed by Gabriel García Márquez, one with its own beguiling set of alternative physical laws.

To start, I asked Don Victor what he calls himself—a healer, a shaman, a medicine man? "I'm not a shaman, that is not an Andean word. I'm not a healer because I don't heal anyone of anything." He called himself a *chakaruna*—a human bridge for people to walk across to get where they need to be. "But a name is just a name," and he suggested the time for names and categories—indeed, for rational thought of any kind—was past.

"In these times people don't need to reason or ask questions so much. That is not the best way to understand the cosmic mind and Mother-Father Earth, which has become so tired from bearing the heavy, dense weight of human thinking, especially in the last two thousand years." He regarded the pandemic as a sign we had fallen away from Mother-Father Earth, that we had lost touch with "our brother and sister animals, plants, minerals, bacteria, and viruses.

"That is why this pause we call the coronavirus is so urgent. It is not a time to analyze or rationalize or to understand. It is a time to replenish and regenerate the absolute energy of the mind."

The man holding forth on my screen was not stern or professorial in the least, but rather jolly. At seventy-one, Don Victor has a genial round face that is remarkably unlined; he wore glasses attached to a cord that formed a vaguely comical loop on either side of his head, and on top of his head, a baseball cap. He was happy to speak "without limitation," which seemed to mean that he would take any question I asked wherever in the world (and some places more distant) he wished to go.

This usually entailed a lengthy excursion that took us far afield, though he always somehow wound his way back to something resembling an answer. When I asked him how he discovered his vocation, he began by warning me that "when we ask one question it automatically has nine answers, and when we want to know what is the answer that will help us, then nine more questions show up, each with another nine answers."

The story of how he found his vocation, for example, begins when Don Victor is five, living alone with his mother in the town of Ayaviri in southern Peru. Every morning at 4:00 a.m. he would slip out of the house and run nine kilometers over three mountains, through waterways and woods, to the tiny Aymara village of Tinajani, where he would arrive at sunrise. Tinajani sits in a dramatic canyon punctuated with intricate red rock formations called Tampu T'oqo that are pierced with caves that are considered sacred. Victor would spend the morning playing in these caves, which he described as "interdimensional portals that hold knowledge of the history of life." The Incans buried their dead in some of these caves, and young Victor would talk to them, not realizing they were spirits. There he met a teacher named Hatun Sonq'o ("Big Heart"), who I *think* but am not 100 percent sure was an actual person. Every day "for three hours he would teach me, and that allowed me to open up my memories of past lives and all the things I can now talk about without limitation." That includes the knowledge that the universe is composed of cosmic vibrations, the lower frequencies associated with anger and violence and limitation, the higher ones with love and peace and gratitude.

What does any of this have to do with Wachuma? Don Victor

was gradually winding his way back to an answer to my question: he works with Wachuma because the plant has the power to raise the frequency of our vibrations.

Risking a follow-up, I asked Don Victor what his mother thought about his predawn adventures. "Mother didn't know. Nobody knew. When I would arrive back in my village, filthy, with my clothes ripped, I would undress and jump in the village water cistern to get clean—I can still feel how cold the water was, since this was in the mountains at thirty-nine hundred meters! When I got out, dripping wet, the whole village could see me. My mother would be furious. She had a braided piece of llama leather with a little ball at the tip and she would hit me with it. But she never knew where I had been."

I asked him about the spirit of the cactus and how it heals people. "It keeps teaching me all the time. I'm certain that one life is not enough to learn everything this plant has to teach us." Don Victor said the plant itself is no more a healer than he is; rather, it is a teacher. We have three bodies, he explained, the physical, the mental, and the spiritual—what he calls "the trinity." (He called each of them a *pacha*—"a world.") "The plant allows all three bodies, little by little, to vibrate at a higher frequency until it is only light, pure light. That is what is meant by illumination." I was lost now, but maybe that was okay: "The plant allows you to disconnect from the mind. You can't figure it out mentally. You need to feel it in your physical body."

Don Victor had his own theory of trauma. "When any part of your body has been affected by destructive energies or trauma, the heart will close down to protect itself. A closed heart will not heal. It will not express its feelings. The mind becomes more active because the heart's not feeling anymore. The mind will go into the past or it will go into the future, which doesn't really exist, and it will get stuck in a chaos, between remembering the past and trying to go into the nonexistent future. And it will lose the gift of life, which is to live and be present in the moment. That is why another word for a gift is a present." Wachuma locates and unblocks the dense energies of trauma so that the mind might quiet and the heart might speak again, returning us to the gift that is the present moment.

Before our time was up, I asked Don Victor for his advice. I told him that I had learned all I could about the plant, how to grow it and how to prepare it, but because of the fires and the pandemic, it seemed unlikely that I could participate in a ceremony, and I was frustrated.

"Two suggestions for you," he began. "There is a way we could do a ceremony online. I would be able to feel your vibrations and specify the proper dosage. This would be my gift." He had apparently done Zoom ceremonies a couple of times with people in Europe. The idea seemed a little weird, and I could tell Taloma was skeptical. It is true that much more of life than we could ever have imagined has migrated to Zoom: classes, meetings, Passover Seders, therapy sessions, funerals, cocktail hours, and on and on. But a medicine ceremony? I wondered about the legal implications—how secure was Zoom?

I asked Don Victor what his second suggestion was.

"The other idea is that you connect profoundly with the spirit of the plant, talk to the plant, and listen to it with your heart. If you have a clear intention and prayer, the plant itself will teach you how much you need to drink and when."

"Solo?" I asked, surprised.

"Sí."

A few days after our session with Don Victor, Taloma, possibly alarmed by his heretical suggestions, proposed a way that we might organize a ceremony after all. We could find an indoor space—a big living room somewhere—and limit the group to six or seven people, so everyone could observe social distancing. We would all get tested a day or two in advance, and Taloma would rejigger certain ritual elements to minimize our risk: separate cups to drink from, separate feathers for smudging, separate everything. And she would invite only scrupulously COVID-conscientious people. This seemed like a reasonable plan; Judith agreed. We scheduled the ceremony for a Saturday night.

The living room where we gathered was a room familiar to me, a place I'd spent time in. Which explains my astonishment when Judith and I arrived early on the appointed Saturday evening: the space had been utterly transformed, the furniture removed and replaced by a large altar crowded with strange and wondrous objects and filling the center of the room. At first glance the room looked like a peasant marketplace in Cuzco, the floor spread with woven cloths in colorful patterns and four large animal skins—a bear, a deer, a bison, and a coyote. On closer inspection, however, every object had been carefully placed in one of four quadrants, each corresponding to one of the cardinal directions and one of the four elements.

Here's a partial list of the objects Taloma had set out on the altar: vials containing purple sand from Big Sur; gigantic seed pods from

Peru; an intricately carved gourd; a bowl of water collected from all over the world; a snake skin; a wooden carving of four grandmothers circled around a lit candle; a marble etched with the seven continents floating in water; a talking stick made from the dried core of a Wachuma cactus; an enormous ear of multicolored corn; fossils; crystals; a dozen or so candles; a Wachuma flower in full bloom; eight stones in the shape of hearts; an abalone shell holding a packet of dried sage leaves; the feathers of a condor and an eagle; a collection of shells; the head of an eagle; and, somewhat incongruously, a photograph of Ruth Bader Ginsburg. Taloma had invited each of us to bring an item to add to the altar. I brought a black faux-barbed-wire bracelet my dad wore in his last years, somewhat inexplicably—it was something Amnesty International sent to contributors.

Taloma wore a white top crossed by a Peruvian sash and a black hat festooned with still more spiritual tchotchkes. She was assisted by "Sam," a lanky thirtyish apprentice with curly black hair and the palest blue eyes. After we took our seats on the floor around the altar, Taloma launched into a lengthy explanation of what was going to happen during the night—the three obligatory cups of Wachuma (plus an optional fourth), the water ceremony at dawn, the option of a tobacco ceremony during the night (more on that in a moment). She spelled out a few rules: no speaking to one another during ceremony; no leaving the circle before dawn except to use the bathroom; no food or water till dawn. Sam handed out buckets to use in case we "got well"—that is, got sick: people occasionally vomited, Taloma explained, but such purging should be regarded as a blessing. Taloma lit a wad of dried sage leaves and, walking slowly among us, wreathed the altar and then each of us in the fragrant smoke. She offered prayers for us and for our troubled country and world. She invoked the spirit of the cactus in teaching us how to heal ourselves and how, once healed, we could better help to heal others. "We are our own best healers," she said. The cactus sees into us, body, mind, and spirit, revealing what needs our attention. Like peyote, it has a penetrating gaze.

There must have been two hours of these preliminaries before Taloma called us up, one by one, for our first cup of Wachuma. When my turn came, Sam poured eight or so ounces into a cup and handed it to Taloma, who said a prayer over the liquid before passing it to me with two hands. I was to silently say my own prayer and then gulp the brown liquid down all at once. The tea was so bitter it sent a shudder up and down my body. Sam now gave me a squirt of Agua de Florida* to rub between my hands, then bring up to my face to inhale. He instructed me to breathe in through the nose and out through the mouth while making a sound—a pattern of breathing we would be encouraged to repeat all through the night, producing a variety of strangely primal sounds in the dark that helped form the ceremony's otherworldly soundtrack. After we had all had our first cup, Taloma began to sing a song about a hummingbird, in a lovely, entrancing voice.

It would be a long, strange night of many elements and episodes. For me, the whole experience was at once more and less powerful than I anticipated. Less powerful because I found the medicine to be remarkably gentle—it never completely took hold of my mind the way the pure mescaline had done, even after I had ingested four cups. There were no visions. What it did was loosen all the cords

*According to Google, Agua de Florida is citrus-scented water used by shamans "to clear heavy energy around the body's energy field" during ceremonies. It also happens to have enough alcohol to serve as a hand sanitizer during a pandemic.

that anchored me to place and time, freeing me to drift along aimlessly on the currents of the evening. But these currents were set in motion less by my own thoughts and emotions than by what was happening in the room: the vibrations of Taloma's singing and Sam's reedy flute; the spooky beating of an owl's wing flapping around my head; the flicker of candlelight on the curved ceiling; and, especially, the shifting emotional register of the audible exhalations, which comprised our sole connection to one another in the dark. These utterances, which seemed to emanate from somewhere deep within us, were by turns plaintive, pained, haunted, reconciled. Together the effect of these sounds was transporting, fostering a mental state that helped me better understand the power of medicine ceremonies, how the chemistry and the shared ritual work together to create a liminal space open to new possibilities. Also, how within that space the group becomes a kind of living, breathing organism, something greater than the sum of the individuals present. I could see—feel—how the medicine softens the edges of self and world in a way that amplifies the power of the ritual by taking us out of ordinary time, and allowing us to suspend disbelief.

This was no small thing. For who were we but a bunch of gringos, most of us white Westerners doing their best to enact an ancient ceremony imported from the Andes. Were we guilty of cultural appropriation? You could say. But such thoughts are the sober disenchantments of daytime; for the duration of that enchanted night they were banished, completely, along with so much else of our current reality. Credit the Wachuma for helping to weave that spell, for making such a ceremony even plausible, but credit also Taloma, who performed her role with absolute conviction. She became for us the medicine carrier, the keeper of ancient wisdom, the *Wachumara*, her words channeling something far beyond the person I had gotten to know. Taloma was in her element, and she was impressive.

My own experience was not at all what I expected. Others had more powerful responses to the medicine, and theirs ended up coloring my own, taking me out of the first person and, strange as this must sound, into the third for much of the night. In retrospect, beside myself was exactly where I needed to be, proposing one possible path out of the nutshell confines of this dismal year.

Soon after we drank our first cup, I began to hear Judith, across the room, crying softly. Taloma went over to work with her, and I could hear them whispering intently. Something had come up for Judith, something she had grappled with in a previous session with another medicine. I had an idea what it was. Her late father had appeared to her, a man she loved dearly but who for most of his life carried a heavy burden of disappointment and fear; he'd been orphaned as a teenager and struggled with his various demons until quite late in life, when he abruptly turned sweet and seemed to find contentment. (A few years before his death, Judith asked him how he could account for the change. He shrugged and told her, "I no longer had time for all that shit, so I let it go.") Judith identified closely with her father and, as she had come to understand, she felt an obligation to carry some of his pain. During the previous journey, she had traveled to the underworld and there met her father, who told her she no longer needed to carry his burden. He released her.

But it was a gift Judith hadn't been able to accept, and this is what I could hear her, barely, whispering about to Taloma. Her mother, still alive, wouldn't allow her to shed any of the weight she was carrying. Judith was herself reluctant to let it go: by now the weight of this inheritance was a part of who she was, integral to her identity and her role in the constellation of the family. What would remain if she managed to let it go? This was a fear she was too fearful to give up.

I could hear Taloma urging Judith to make a move, renounce her inheritance. "It's your choice. We make the world with our words. Say it. Say the words right now." But Judith, crying more loudly now, couldn't bring herself to say the words. This was painful to hear—or rather, *not* hear. I felt helpless, unable to offer any words or a touch of comfort. Judith must have been reading my mind, because I heard her whisper to me across the room: "I need to do this myself." Whatever effect the medicine had had on me to this point now disappeared.

Taloma offered Judith a tobacco ceremony, something I knew about, having endured one a few weeks prior, when I was getting to know Taloma. Sorry to introduce another plant medicine at this point in our story, but it is common in Indigenous ceremonies for healers to deploy more than one. I had been surprised to read that many shamans regard tobacco as the most powerful of all plant medicines, and it figures prominently in ceremonies in many traditions, including the Native American peyote meeting. Westerners today bring a lot of negative attitudes to tobacco, regarding the plant as irredeemably evil, but, as Taloma explained, that is only because white people had abused and exploited this sacred plant when they arrived in the Americas, transforming it from a sacred medicine into a lethal and addictive habit.

There are a few different ways tobacco is used in Indigenous ceremonies, but usually as a means of purging evil or destructive energies. In Taloma's version, the recipient stands before her and closes one nostril while she offers a brief prayer that ends with the words "body, mind, and spirit." On the word "spirit," you inhale deeply while Taloma, using a syringe, shoots tobacco juice deep into your sinus cavity. A wave of fire races across the top of your skull from front to back and then travels down your spine. It is a bracing sensation. Taloma encourages you to stomp your feet, shake out your arms, move your hips, vocalize with abandon, and let go of whatever emotions you are holding. After the firestorm subsides, your mind feels freshly scrubbed and, at least for a while, cleared and wonderfully calm.

It wasn't until after we had all drunk our third cup of Wachuma that Judith asked Taloma for the tobacco. Judith is ordinarily an extremely private person, so doing such a thing in a group took courage. I had a tip to offer, but felt constrained by Taloma's no-talking rule. I waited until she stepped out of the room to prepare the medicine and then stage-whispered to Judith: "Whatever you do, don't swallow!" I had let some of the tobacco juice slide down the back of my throat and spent an uncomfortable night feeling like I had swallowed the contents of an ashtray.

The tobacco ceremony was not easy to watch. Now that I was completely sober, my prayers turned to Judith, as did the thoughts of everyone in the room. She seemed completely unselfconscious; I wondered if our collective energies had buoyed her. We watched from our respective corners of the room as, on the word "spirit," the medicine coursed through her body, seizing control of her arms and legs and vocal cords, all helpless before its force. Deep guttural animal sounds emerged from her throat while her body, seemingly possessed, launched into a kind of spastic dance. Sam sang a song about a condor, chorus after chorus, while Taloma moved in rhythm with Judith's swaying, working her hands over her body (so much for social distancing) and ritualistically yanking out knots of bad juju from her belly, her neck, and the top of her head.

The whole ceremony lasted only a few minutes, and when the storm subsided, Judith seemed becalmed. She told me later she felt good, emptied out and cleansed. Something had shifted in her; whether it would last remained to be seen.

I felt as though we had witnessed a kind of faith healing, and it helped me understand the power of doing this sort of work in a group. For in addition to the medicine and the rituals, there were the pooled energies of other people, all of it trained on one person, one outcome. We had also witnessed how, three cups in, the Wachuma could relax one's mental and physical defenses (Judith is someone who ordinarily can't even tolerate a massage!), softening the grip of the rigid narratives we tell ourselves about who we are and have to be. With the help of the medicine, Judith had put something supposedly core and unshakable about herself up for grabs. Although there was no guarantee this would happen, a space had been created in which a new story might begin to take shape.

After all the drama, I was eager to return to my own reveries, so I asked for the discretionary fourth cup. When I came up to Taloma's altar, she asked a few questions to assess my state of mind and agreed I should have more. She decided to strengthen this cup by adding a big spoonful of powdered Wachuma from Peru. This thickened the brew, making it even more difficult to swallow, but I was grateful for the speed with which it returned me to my inward journey, sending me further and deeper than before.

I spent the rest of the night carried along on warm waters of thought and feeling, in the kind of agreeably drifty meditation that often follows the climax of a psychedelic experience, though the climax hadn't been mine. I visited with people in my life both alive and departed. Some knotty issues I had planned to work on no longer seemed knotted; they passed into my field of awareness and then passed out, not so much resolved as released. At one point I wondered why *I* wasn't having an emotional or spiritual crisis, whether my defenses were too strong for the medicine to breach or if there just wasn't as much going on in my unconscious as I liked to think.

Eventually I turned my attention to the exercises Taloma had suggested we work with—the "three levels of forgiveness" and the practice of gratitude, exercises Don Victor had also talked about. By asking forgiveness for the pain we have caused others, Taloma had told us, "we cut our cords to the discordant or destructive energies that connect us to others in the past." Next we offer forgiveness to those who have caused us to suffer. I summoned my father, Judith, my son, sisters, certain friends, and asked for and offered these words. As it is, the medicine attenuates the bonds of the past, making it easier to let go of regrets. And then we forgive ourselves.

What follows forgiveness is gratitude, which I now felt break over me in a warm wave of tears—gratitude for the gift of having these people in my life, for having this life and however many more years of it remained, and for having been introduced to a plant with the power to summon these tears and help me to see, even in this bleak, bleak season of loss, just how much I had to be grateful for. Despair no longer felt like an option.

(How saccharine these words must sound! I can only imagine. I'm afraid banality is an unavoidable hazard of working with psychedelics; they are profound teachers of the obvious. But sometimes those are exactly the lessons we need.)

I was still drifting on these warm currents of emotion when

Taloma began to close the ceremony with a water prayer. We hadn't had a sip of water all night, and the prospect of drinking some now was sweet. But first the ceremony. Taloma lit a fat roll of tobacco, blew some smoke over the pitcher of water, and offered a long, plaintive prayer of "gratitude for the sacred water" that moved in widening gyres from the purity of this life-giving water she had drawn from the springs at Esalen, to the fouling of the earth's rivers and seas by humanity's carelessness and greed, to the even larger desecrations of nature in our time, the corruption of our country, and the proximate specters of the virus and fires. The pandemic and the great pause it had forced upon the world was the opportunity, she fervently prayed, for humanity to awaken to what we had done to the earth and change the way we live upon it. She reminded us the lockdown had shown how quickly nature could heal herself if given the chance. "But the time is *now*," she said, her voice cracking under the pressure of an urgency she seemed to be channeling from the depths of the earth itself. Could this be our last chance?

The water prayer took me by surprise. Without warning, Taloma had shocked us out of our nighttime reveries and back into the daylight of history, recalling us to the perils waiting outside the space and time we had had the privilege of sharing overnight. What had been a time out of time, a brief, blessed respite from the fires and the virus, was now over. What came next? Taloma spoke of the ripples in water and how far they could travel. She prayed for us to become ripples of healing, traveling out from this room to repair the world before it was too late. To feel the raw force of her words you would probably have to be there, to have had your heart opened up by this plant, but they were as gutting as they were beautiful.

As the first soft light of the new day crept into the room, we greedily drank the pure water and gave thanks for it.

The ceremony's last act was the passing of the talking stick, an opportunity for each of us to share what had happened overnight, and to try to make some sense of it. I was struck by how strongly Judith's experience had inflected everyone else's, especially how it had brought the spirits of our parents into the space we shared; mothers loomed large in the accounts several of us offered. Our separate psyches hadn't merged, by any means, but they had overlapped, and how long had it been since anything like that had happened?

When Judith took the stick, she sheepishly apologized for "all the drama last night." And then she pronounced the words she hadn't been able to say before, that she was ready to put down her father's burden. Yet she did it in the future tense. When Taloma pointed this out, reminding her that "the future doesn't exist," Judith repeated them, now in the present tense, and smiled.

Before everyone dispersed to return to their lives, we took a selfie of our group, squeezing together to fit in the frame as if in a dream in which the pandemic was over. In the picture all of us look ragged and exhausted yet buoyant, too, and connected to one another in a way that we hadn't been a dozen or so hours before. It was as if we had gone down a river together on a raft, endured some sort of ordeal we couldn't quite describe but sensed had left us changed, in ways that Taloma said might take days or weeks to recognize. "The spirit of the plant is now inside of you forever," she told us. "You can call upon it at any time." After packing up her altar, returning the sacred objects to their woven bags and wooden boxes, Taloma handed Judith the Wachuma blossom, faded now but still gorgeous.

Acknowledgments

To thank everyone who contributed in one way or another to the research, writing, and publication of *This Is Your Mind on Plants* means going back more than twenty-five years. That's when my friend and editor at *Harper's Magazine*, Paul Tough, sent me a copy of *Opium for the Masses*, the underground press book that launched my brief career as an opium grower and inspired the original version of this book's chapter on that plant. I also owe a large debt of gratitude to the publisher of *Harper's Magazine*, then and now, John R. "Rick" MacArthur. Rick went above and beyond what any normal publisher would do in order to make it possible (and safe) for me to publish that piece; thanks, too, to Lewis Lapham, the editor of *Harper's* at that time, for commissioning it and for supporting my earliest efforts to write about the doings in my garden. Victor Kovner, the venerable First Amendment lawyer, played a critical role in helping that piece see the light of day. So did my brother-in-law, ace attorney Mitchell Stern, who helped me see straight and stay calm through the whole ordeal. And even though I ultimately didn't take his advice, I'm grateful to criminal defense attorney David Atkins for his counsel and care.

ACKNOWLEDGMENTS

An earlier, shorter version of the chapter on caffeine first appeared as an audiobook published by Audible in 2020. I'm grateful to the team at Audible, but particularly to Doug Stumpf for deeming the idea promising enough to commission, and to Susan Banta for her scrupulous fact-checking and copyediting of the manuscript. I've added a considerable amount of new material on tea, a subject about which I've learned much over the years from Sebastian Beckwith, the proprietor of In Pursuit of Tea. David Hoffman, the pioneering tea hunter, importer, and collector, was also generous with his passion and boundless knowledge, as well as a memorable tasting. Thanks to my friend and colleague Peter Sacks, for reminding me of the role of caffeine in "The Rape of the Lock," and to Raj Patel for pointing me to readings about the political economy of tea and coffee that I would never have found on my own.

The debts I incurred reporting on mescaline are numerous. Early on, Adele Getty and Michael Williams of the Limina Foundation taught me a great deal about mescaline as it has been used in both Indigenous and Western contexts. Thanks to my friend Cody Swift, founder of the Indigenous Peyote Conservation Initiative, and his colleague Miriam Volat for educating me about the threat to the peyote cactus and for introducing me to several of the members of the Native American Church who appear in the narrative. IPCI's work conserving peyote for Native Americans is urgent and deserves our support (ipci.life). Jerry Patchen, an attorney who has been fighting since the 1990s for Native Americans' right to use peyote, provided rich insight as well as some illuminating historical documents. Adrian Jawort read the chapter with care, bringing the eye of a Native American to my account of peyotism. I'm grateful to Nick Cozzi and Dave Nichols for educating me on the chemistry and pharmacology of mescaline. Keeper Trout and Tania Manning tutored me on the bewildering botany of the cacti lumped together under the rubric San Pedro; Martin Terry did the same for the botany of peyote. I owe a debt to Michael Zeigler, one of the wise men of this community, for his long perspectives and horticultural generosity. Bob Hass helped me to understand the haiku consciousness that mescaline sponsors, in my mind if nowhere else. And thanks to Bob Jesse, Joe Green, Mike Jay, Bia Labate, Françoise Bourzat, Tom Pinkson, Dawn Hofberg, and Erika Gagnon for deepening my knowledge of this plant medicine and its history. Finally, I always feel better publishing something after Bridget Huber has fact-checked it with her fine-tooth comb, and I sleep much better after my old friend Howard Sobel and his colleague Rob Ellison, of Latham & Watkins, have read it with a keen legal eye; thank you, Howard and Rob.

As ever, I'm grateful to the one and only book editor I've ever worked with, Ann Godoff, for her enthusiasm and sure-footed guidance of this project, as well as to the one and only literary agent I have ever had, Amanda Urban. Each new upheaval in the publishing industry serves to remind me of how very fortunate I have been to have these two wise women in my corner for the whole of my career. Their respective teams are the best in the business. Special thanks to Sarah Hutson, Casey Denis, Sam Mitchell, Darren Haggar, Karen Mayer, Danielle Plafsky, John Jusino, and Diane McKiernan at Penguin Press; at ICM, Jennifer Simpson, Sam Fox, Rory Walsh, and Ron Bernstein; and, at Curtis Brown in London, Daisy Meyrick and Charlie Tooke. And a shout-out to Simon Winder at Penguin UK for his sharp editing, years of support, and salutary reminders that not all readers are American. It is a privilege and a pleasure to work with all of you.

There is a third wise woman who has played a critical role in every one of my books, both behind the scenes and, this time, *in* several scenes, and that of course is Judith Belzer, my wife and life partner. Thank you for the sounding board, the superb advice, the deft editing, the integration sessions, *and* your willingness to come along on this ride and share your experience: you have been generous beyond what I could reasonably expect. Thank you, too, Isaac Pollan, for your continuing interest in and support for your dad's journalistic adventures; I always get something out of our conversations about the work, not to mention good advice about the optimal way to brew coffee. I'm endlessly grateful for my writer pals, for the shoptalk and the counsel: Mark Edmundson, Mark Danner, Gerry Marzorati, Jack Hitt, and Dacher Keltner, dear friends all. Whether on the trail or the phone, you make the work we do so much less lonely.

And last but hardly least in my gratitude, since they float the whole enterprise, are the readers. Some of you have been with me as far back as 1991, when I published *Second Nature*. I feel lucky to have found a community of readers willing to come along with me on this improbable, winding journey, from the garden to the farm and kitchen and then to the mind, and, now, back to where we started, with the plants we rely on and the human desires on which they so cleverly play. Thanks for your open-mindedness, curiosity, generosity, and, especially, all your letters, emails, posts, and tweets—I learn easily as much from you as you do from me. I count it a privilege every time you grant me a few hours of your time and attention.

Selected Bibliography

Opium

Baum, Dan. "Legalize It All." *Harper's Magazine*, April 2016.

———. *Smoke and Mirrors: The War on Drugs and the Politics of Failure.* New York: Little Brown, 1996.

Booth, Martin. *Opium: A History.* New York: Thomas Dunne Books, 1998.

De Quincey, Thomas. *Confessions of an English Opium-Eater.* Norwalk, CT: Easton Press, 1978.

Halpern, John H., MD, and David Blistein. *Opium: How an Ancient Flower Shaped and Poisoned Our World.* New York: Hachette, 2019.

Hogshire, Jim. *Opium for the Masses: Harvesting Nature's Best Pain Medication.* Port Townsend, WA: Loompanics Unlimited, 1994.

Keefe, Patrick Radden. "The Family That Built an Empire of Pain." *New Yorker,* October 23, 2017.

Lenson, David. *On Drugs.* Minneapolis: University of Minnesota Press, 1995.

Macy, Beth. *Dopesick: Dealers, Doctors and the Drug Company That Addicted America.* New York: Back Bay Books, 2019.

Nutt, David. *Drugs Without the Hot Air: Minimising the Harms of Legal and Illegal Drugs.* Cambridge, UK: UIT Cambridge, 2012.

Pendell, Dale. *Pharmako/Poeia: Power Plants, Poisons, and Herbcraft.* Berkeley: North Atlantic Books, 2010.

Caffeine

Allen, Stewart Lee. *The Devil's Cup: A History of the World According to Coffee*. New York: Soho Press, 1999.

Balzac, Honoré de. *Treatise on Modern Stimulants*. Translated by Kassy Hayden. Cambridge, MA: Wakefield Press, 2018.

Braudel, Fernand. *The Structures of Everyday Life, Vol. 1*. New York: Harper and Row, 1981.

Carpenter, Murray. *Caffeinated: How Our Daily Habit Helps, Hurts, and Hooks Us*. New York: Plume, 2015.

Couvillon, Margaret J., et al. "Caffeinated Forage Tricks Honeybees into Increasing Foraging and Recruitment Behaviors." *Current Biology* 25, no. 21 (November 2, 2015): 2815–18. doi:10.1016/j.cub.2015.08.052.

Ekirch, A. Roger. *At Day's Close: Night in Times Past*. New York: W. W. Norton, 2005.

Grosso, Guissepe, et al. "Coffee, Caffeine and Health Outcomes: An Umbrella Review." *Annual Review of Nutrition* 37 (2017): 131–56.

Halprin, Mark. *Memoir from Antproof Case*. New York: Harcourt, Brace, 1995.

Hobhouse, Henry. *Seeds of Change: Six Plants That Transformed Mankind*. Berkeley: Counterpoint, 2005.

Hohenegger, Beatrice. *Liquid Jade: The Story of Tea from East to West*. New York: St. Matin's Press, 2006.

Houtman, Jasper. *The Coffee Visionary: The Life and Legacy of Alfred Peet*. Mountain View, CA: Roundtree Press, 2018.

Juliano, Laura M., Sergi Ferré, and Roland R. Griffiths, "The Pharmacology of Caffeine." *The ASAM Principles of Addiction Medicine: Fifth Edition*. Wolters Kluwer Health Adis (ESP), 2014.

Kretschmar, Josef A., and Thomas W. Baumann. "Caffeine in Citrus Flowers." *Phytochemistry* 52, no. 1 (September 1999): 19–23. doi:10.1016/S0031-9422(99)00119-3.

Kummer, Corby. *The Joy of Coffee*. Boston: Houghton Mifflin, 1995.

Milham, Willis I. *Time and Timekeepers: Including the History, Construction, Care, and Accuracy of Clocks and Watches*. New York: Macmillan, 1923.

Mintz, Sidney W. *Sweetness and Power: The Place of Sugar in Modern History*. New York: Penguin, 1985.

Morris, Jonathan. *Coffee: A Global History*. London: Reaktion Books, 2019.

Pendell, Dale. *Pharmako/Dynamis: Stimulating Plants, Potions, and Herbcraft*. San Francisco: Mercury House, 2002.

Pendergrast, Mark. *Uncommon Grounds: The History of Coffee and How It Changed Our World*. New York: Basic Books, 1999.

Reich, Anna. "Coffee and Tea: History in a Cup." *The Herbarist Archives* 76 (2010).

Reid, T. R. "Caffeine—What's the Buzz?" *National Geographic Magazine*. January 2005.

Saberi, Helen. *Tea: A Global History*. London: Reaktion Books, 2010.

Schivelbusch, Wolfgang. *Tastes of Paradise: A Social History of Spices, Stimulants, and Intoxicants*. Translated by David Jacobson. New York: Pantheon Books, 1992.

Sedgewick, Augustine. *Coffeeland: One Man's Dark Empire and the Making of Our Favorite Drug*. New York: Penguin Press, 2020.

Spiller, Gene A., ed. *Caffeine*. Boca Raton, FL: CRC Press, 1998.

Standage, Tom. *A History of the World in 6 Glasses*. New York: Bloomsbury, 2005.

Ukers, William H. *All About Coffee*. New York: The Tea and Coffee Trade Journal Company, 1922.

van Driem, George. *The Tale of Tea: A Comprehensive History of Tea from Prehistoric Times to the Present Day*. Leiden, NL: Brill, 2019.

Walker, Matthew. *Why We Sleep: Unlocking the Power of Sleep and Dreams*. New York: Scribner, 2017.

Weinberg, Alan, and Bonnie K. Bealer. *The World of Caffeine: The Science and Culture of the World's Most Popular Drug*. Abingdon, UK: Routledge, 2001.

Wright, G. A., et al. "Caffeine in Floral Nectar Enhances a Pollinator's Memory of Reward." *Science* 339, no. 6124 (March 8, 2013): 1202–4. doi:10.1126/science.1228806.

Mescaline

Artaud, Antonin. *Antonin Artaud: Selected Writings*. Edited by Susan Sontag and translated by Helen Weaver. New York: Farrar, Straus and Giroux, 1976.

Bourzat, Françoise, and Kristina Hunter. *Consciousness Medicine: Indigenous Wisdom, Entheogens, and Expanded States of Consciousness for Healing and Growth: A Practitioner's Guide*. Berkeley: North Atlantic Books, 2019.

Brown, Dee. *Bury My Heart at Wounded Knee: An Indian History of the American West*. New York: Holt, Rinehart and Winston, 1970.

Calabrese, Joseph D. *A Different Medicine: Postcolonial Healing in the Native American Church*. New York: Oxford University Press, 2013.

Gwynne, S. C. *Empire of the Summer Moon: Quanah Parker and the Rise and Fall of the Comanches, the Most Powerful Indian Tribe in American History*. New York: Scribner, 2011.

Hass, Robert, ed. *The Essential Haiku: Versions of Bashō, Buson, and Issa*. New York: Ecco Press, 1995.

Huxley, Aldous. *The Doors of Perception*. New York: Harper and Row, 1954.

Jay, Mike. *Mescaline: A Global History of the First Psychedelic*. New Haven: Yale University Press, 2021.

Jesse, Bob. *On Nomenclature for the Class of Mescaline-Like Substances and Why It Matters*. San Francisco: Council on Spiritual Practices, 2000.

Keeper Trout and Friends, ed. *Trout's Notes on San Pedro and Related* Trichocereus *Species: A Guide to Assist in Their Visual Recognition; with Notes on Botany, Chemistry, and History*. Austin, TX: Mydriatic Productions/Better Days Publishing, 2005.

LaBarre, Weston. *The Peyote Cult*. Norman: University of Oklahoma Press, 1989.

Lame Dog. *Seeker of Visions*. New York: New York University Press, 1976.

Maroukis, Thomas C. *Peyote Road: Religious Freedom and the Native American Church*. Norman: University of Oklahoma Press, 2012.

Pendell, Dale. *Pharmako/Gnosis: Plant Teachers and the Poison Path*. Berkeley: North Atlantic Books, 2010.

Pinkson, Tom Soloway. *The Shamanic Wisdom of the Huichol: Medicine Teachings for Modern Times*. Rochester, VT: Destiny Books, 2010.

Shulgin, Alexander T., and Ann Shulgin. *PiHKAL: A Chemical Love Story*. Berkeley: Transform Press, 1991.

Smith, Huston, and Reuben Snake. *One Nation Under God: The Triumph of the Native American Church*. Santa Fe, NM: Clear Light Publishers, 1996.

Stewart, Omer C. *Peyote Religion: A History*. Norman: University of Oklahoma Press, 1993.

Swan, Daniel C. *Peyote Religious Art: Symbols of Faith and Belief*. Jackson: University Press of Mississippi, 1999.

Index

INDEX

ALSO AVAILABLE

FOOD RULES

An Eater's Manual

Eating doesn't have to be so complicated. In this age of ever-more-elaborate diets and conflicting health advice, *Food Rules* brings welcome simplicity to our daily decisions about food. Written with clarity and wit, this easy-to-use handbook lays out a set of straightforward, memorable rules for eating wisely, one per page, accompanied by a concise explanation, drawing from a variety of traditions. No matter the setting, this is the perfect guide for anyone who ever wondered, "What should I eat?"

COOKED

A Natural History of Transformation

In *Cooked*, Michael Pollan explores the previously uncharted territory of his own kitchen. Here he discovers the enduring power of the four classical elements—fire, water, air, and earth—to transform the stuff of nature into delicious things to eat and drink. Apprenticing himself to a succession of culinary masters, Pollan learns how to grill with fire, cook with liquid, bake bread, and ferment everything from cheese to beer. The lessons move beyond the practical to become an investigation of how cooking connects us. Now a series on Netflix.

HOW TO CHANGE YOUR MIND

What the New Science of Psychedelics Teaches Us About Consciousness, Dying, Addiction, Depression, and Transcendence

A brilliant and brave investigation into the medical and scientific revolution taking place around psychedelic drugs—and the spellbinding story of Michael Pollan's own life-changing psychedelic experiences. *How to Change Your Mind* is both dazzling and edifying, taking us into an unexpected new frontier in our understanding of the mind, the self, and our place in the world.